J. WESTON
WALCH
PUBLISHER

topShelf
PRECALCULUS

$$y = \log_b x$$

10 20

J. Bryan Sullivan

User's Guide
to
Walch Reproducible Books

1 2 3 4 5 6 7 8 9 10

ISBN 0-8251-4620-8

J. Weston Walch, Publisher
P.O. Box 658 • Portland, Maine 04104-0658
walch.com

Printed in the United States of America

Contents

To the Teacher

Creative problem solving, precise reasoning, effective communication, and alertness to the reasonableness of results are some of the essential areas of mathematics that educators have specified as necessary in the development of all students to function effectively in this century. It is in the spirit of the aforementioned competencies that *Top Shelf Math* is offered to teachers of mathematics.

Top Shelf Math is intended to help students become better problem solvers. The problems presented in this format are challenging, interesting, and can easily be blended into the teaching styles and strategies of teachers who are seeking supplementary problems that support and enhance the curriculum being taught.

In general, the topics selected for each of the major content areas of *Top Shelf Math* are typical of those found in the curricula of similarly named courses offered at the high school and early college level. All of the problems presented can be used by the teacher to help his or her students improve their problem-solving skills without slowing the pace of the course in which the students are enrolled.

Other areas of utilization of the problems presented in *Top Shelf Math* could be by teachers to help prepare their students for achievement tests and advanced-placement tests. Math-team coaches will find these problems especially useful as they prepare their students for math competitions. It is recommended that students save the problems and solutions presented in *Top Shelf Math* because they provide a rich resource of mathematical skills and strategies that will be useful for preparing to take standardized tests and enrollment in future mathematics courses.

Mathematical thought along with the notion of problem solving is playing an increasingly important role in nearly all phases of human endeavor. The problems presented in *Top Shelf Math* help provide the teacher with a mechanism for the students to witness a variety of applications in a wide sphere of real-life settings.

In problems requiring a calculator solution, it is recommended that only College Board-approved calculators be used. In addition, some problems will suggest that a calculator not be used and that the solution will require an algebraic procedure.

The approach for solving the problems presented in *Top Shelf Math* is consistent with emphasis by national mathematics organizations for reform in mathematics teaching and learning, content, and application by taking advantage of today's technological tools that are available to most if not all high school and college students enrolled in similarly named courses. *Top Shelf Math* provides the teacher with a balance of using these tools as well as well-established approaches to problem solving.

We hope that you will find the problems useful as general information as well as in preparation for higher-level coursework and testing. For additional books in the *Top Shelf Math* series, visit our web site at walch.com.

Exponents and Radicals

Exponents and radicals are basic to all areas of high school mathematics. You learn the rules in algebra I and then apply them in many different problems in future math courses.

Many students misuse the first two rules for using exponents.

Rules for Using Exponents

a and $b \neq 0$ and p and r are positive integers.

1. $a^p \cdot a^r = a^{p+r}$, such as $2^2 \cdot 2^r = 2^{2+r}$
2. $(a^p)^r = a^{pr}$

Many students misuse these two rules. A suggestion—think of rule 2 as multiplication, $^{p})^{r}$:

3. $a^{-p} = \frac{1}{a^p}$
4. $(ab)^p = a^p b^p$
5. $\left(\frac{a}{b}\right)^p = \frac{a^p}{b^p}$
6. $a^0 = 1$
7. $\frac{a^p}{a^r} = a^{p-r} = \frac{1}{a^{r-p}}$, if $p > r$

Radicals: $\sqrt[r]{a}$ is the radical, a is the radicand, and r is the root.

The principal root is the positive root.

8. $\sqrt[r]{a^p} = a^{\frac{p}{r}}$; you can remember this as p means power and r means root.
9. $\sqrt[pr]{a} = \sqrt[p]{\sqrt[r]{a}} = \sqrt[r]{\sqrt[p]{a}} = a^{\frac{1}{pr}}$, such as $\sqrt[8]{36} = \sqrt[4]{\sqrt{36}} = \sqrt[4]{6}$
10. $\sqrt[r]{a^p} = (\sqrt[r]{a})^p$, such as $\sqrt[5]{(32)^2} = (\sqrt[5]{32})^2 = 2^2 = 4$

Example 1

Simplify without radicals: $\sqrt[5]{\dfrac{\sqrt[4]{5^3 \cdot a^5}}{\sqrt[3]{5^2 \cdot b^4}}}$

$$\sqrt[5]{\frac{\sqrt[4]{5^3 \cdot a^5}}{\sqrt[3]{5^2 \cdot b^4}}} = \left(\frac{(5^3 \cdot a^5)^{\frac{1}{4}}}{(5^2 \cdot b^4)^{\frac{1}{3}}}\right)^{\frac{1}{5}} = \frac{\left[(5^3 \cdot a^5)^{\frac{1}{4}}\right]^{\frac{1}{5}}}{\left[(5^2 \cdot b^4)^{\frac{1}{3}}\right]^{\frac{1}{5}}} =$$

$$\frac{(5^3 \cdot a^5)^{\frac{1}{20}}}{(5^2 \cdot b^4)^{\frac{1}{15}}} = \frac{5^{\frac{3}{20}} \cdot a^{\frac{5}{20}}}{5^{\frac{2}{15}} \cdot b^{\frac{4}{15}}} = \frac{5^{\frac{3}{20} - \frac{2}{15}} \cdot a^{\frac{1}{4}}}{b^{\frac{4}{15}}} = \frac{5^{\frac{1}{60}} \cdot a^{\frac{1}{4}}}{b^{\frac{4}{15}}}$$

TRY IT Practice Activities

1. Find all positive values of x that satisfy $x \cdot \sqrt[x]{x^2} = x^x$.

2. If $3 + \sqrt{b}$ is equivalent to the reciprocal of its conjugate, what is the value of b?

3. If a and b are positive and $\sqrt{(a+2)(b+3)} = \sqrt{2b} + \sqrt{3a}$, find the product ab.

4. Find $p + q$ in terms of a, b, and c, where $a > b$.

   $$\left(\frac{x^b y^a x^a}{y^b}\right)^c = x^p y^q$$

5. For the expression $\dfrac{\sqrt[3]{9} \cdot \sqrt[4]{4}}{\sqrt[6]{12}} = \sqrt[m]{a}$, where a is a whole number and m is the smallest possible root, find the sum $a + m$.

6. Simplify: $\dfrac{\left(9^x \cdot 3^2 \cdot \dfrac{1}{3^{-x}}\right) - 27^x}{3^{3x} \cdot 9}$

7. The square of $a + b\sqrt{3}$ is $84 - 30\sqrt{3}$. Find all ordered pairs of (a, b), where a and b are integers.

8. Simplify: $\dfrac{\sqrt{2}}{\sqrt{2} + \sqrt{3} - \sqrt{5}}$

9. Find all real value(s) of x for which the following is true:

   $$\frac{x + \sqrt{x+1}}{x - \sqrt{x+1}} = \frac{5}{11}$$

Polynomials and Factoring

This unit deals with polynomials and factoring. The material learned in this unit will be used in many of the other units.

Polynomials are written in descending order.

A polynomial is an algebraic expression that can be written in the form: $a_nx^n + a_{n-1}x^{n-1} + a_{n-2}x^{n-2} + \ldots + a_3x^3 + a_2x^2 + a_1x^1 + a_0$, where $a_0, a_1, \ldots, a_{n-1}, a_n$ are numbers or constants or coefficients and n is a nonnegative integer.

As we know, polynomials are written in descending order, such as $5x^4 + x^2 - 4x + 2$.

The coefficient of the highest power is called the **leading coefficient,** and the term without a variable is called the constant term.

The **degree of a polynomial** is the number from the highest exponent. The degree of the polynomial above is 4, the leading coefficient is 5, and the constant term is 2. A polynomial of degree 1 is a linear polynomial $(2x + 3)$, one of degree 2 is a **quadratic** $(x^2 - 3x + 1)$, and one of degree 3 is a cubic $(2x^3 + 4x - 5)$.

Factoring

If a polynomial Q is a factor in every polynomial $P_1 \ldots P_n$, then Q is considered a common factor or common divisor. A polynomial that can't be factored is a prime polynomial. The first rule of factoring is to factor the greatest common factor (GCF).

Factoring rules

1. $a^2 - b^2 = (a + b)(a - b)$ and is called difference of two squares.
2. $a^2 \pm 2ab + b^2 = (a \pm b)^2$ and is called a perfect trinomial square.
3. $a^3 \pm b^3 = (a \pm b)(a^2 \pm ab + b^2)$ and is called sum/difference of two cubes.

4. $abx^2 + (ad + bc)x + cd = (ax + c)(bx + d)$; this is the general type, cd is the constant term, the coefficient of x^2 is ab, and the coefficient of x is $(ad + bc)$.

The most difficult problems to factor are those with more than three terms.

The most difficult problems to factor are those with more than three terms. The method used is called grouping, and there are two methods.

Example 1

Factor: $m^2 - 4x^2 - 4x - 1$.

Solution:
Group the last three terms, $m^2 - (4x^2 + 4x + 1) =$
$m^2 - (2x + 1)^2 = (m + 2x + 1)(m - (2x + 1))$. The answer is
$(m + 2x + 1)(m - 2x - 1)$.

We used two of the rules—difference of two squares and perfect trinomial.

Example 2

Factor: $x^2 - y^2 + 4y - 4x$.

Solution:
Group in pairs, $(x^2 - y^2) + (4y - 4x) = (x - y)(x + y) + 4(y - x)$; you need a common factor from each pair, so factor out a –4.

$\therefore (x - y)(x + y) - 4(x - y)$; $(x - y)$ is a common factor, so the answer is $(x - y)(x + y - 4)$.

You can check to see that you factored correctly by multiplying your answer to see if you obtain the original problem.

π TRY IT Practice Activities

1. Write as a product of prime factors: $(a + b)(a - b) - c(c + 2b)$.

2. Factor completely: $4a^2 - 9b^2 + 20a + 24 - 6b$.

3. Find the sum of the factors: $4a^2 + 4ab + b^2 - 2a - b - 12$.

4. Find the least common multiple of A, B, and C, where

   $A = 12x^5 + 24x^4 + 12x^3$; $B = 15x^4 - 15x^2$; $C = 9x^6 - 18x^5 + 9x^4$.

5. Factor completely: $yx^2 - x^2z + y^2z - yz^2 + xy^2 - xz^2$.

6. Factor completely: $x^2 - 2xy + y^2 + 10x - 10y + 9$.

7. Factor completely: $16a^{x+2} - 20a^{x+3} + 4a^{x+4}$.

8. Factor completely: $x^4 + (3 + t)x^2 - 2t - 10$.

Rational Expressions

This unit deals with both rational expressions and equations. This unit will strengthen the algebraic skills that you will need for the other units.

Make sure you check your answers for rational equations so there are no zeros in any denominator.

A **rational algebraic expression** is a sum, difference, product, or quotient of terms of the form $\frac{p(x)}{q(x)}$, where $p(x)$ and $q(x)$ are polynomials and $q(x) \neq 0$. The definition of a polynomial is given in the unit "Polynomials and Factoring." It is suggested that you reread the introduction. You can perform all four operations (+, –, ×, ÷) on rational expressions. These operations require factoring to obtain common denominators and to simplify the final answer. You need to be very skillful and careful when working on these problems. Make sure you check your answers for rational equations so there are no zeros in any denominator.

Rules for operating on rational expressions:

1. $\frac{a(x) \cdot b(x)}{b(x) \cdot c(x)} = \frac{a(x)}{c(x)}$; cancel the like terms, the $b(x)$'s.

2. $\frac{a(x)}{c(x)} + \frac{b(x)}{d(x)} = \frac{a(x) \cdot d(x) + b(x) \cdot c(x)}{c(x) \cdot d(x)}$; get a common denominator and combine the terms in the numerator. Don't lose the denominator.

3. $\frac{a(x)}{c(x)} - \frac{b(x)}{d(x)} = \frac{a(x) \cdot d(x) - b(x) \cdot c(x)}{c(x) \cdot d(x)}$; same as rule 2, but subtract and watch the signs.

4. $\frac{a(x)}{b(x)} \cdot \frac{c(x)}{d(x)} = \frac{a(x) \cdot c(x)}{b(x) \cdot d(x)}$; multiply the terms in the numerator and denominator if you can't cancel.

5. $\dfrac{\frac{a(x)}{b(x)}}{\frac{c(x)}{d(x)}} = \frac{a(x)}{b(x)} \div \frac{c(x)}{d(x)} = \frac{a(x)}{b(x)} \cdot \frac{d(x)}{c(x)}$

   Divide or invert and multiply.

TRY IT Practice Activities

1. For what integer value(s) of x does $\frac{(35 - 17x + 2x^2)}{(25 - x^2)}$ represent a perfect square of some rational number?

2. If $\frac{(n-1) + n + (n+1)}{n} = \frac{2{,}002 + x + 2{,}004}{x}$ and $n \neq 0$, find x.

3. If $\frac{x^2 + 3xy - 4y^2}{x^2 - y^2} = \frac{13}{7}$, $x \neq y$, find the ratio of y to x.

4. Let $f(x) = x^2$ and $g(x) = 3x + 10$, simplify the following expression: $\frac{f(x+3) - g(x+3)}{(x+3) + 2}$.

5. Solve for x: $\frac{\frac{x}{x+1} - \frac{1-x}{x}}{\frac{x}{1+x} + \frac{1-x}{x}} = 1$.

6. If $A + B = 10$ and $A - B = 6$, what is the value of $\frac{(3A^2 + 6AB + 3B^2)}{(5A^2 - 5B^2)}$?

7. For how many integral value(s) of x does $\frac{(3x + 17)}{(x + 2)}$ have an integral value?

8. Find the value(s) of constants C and D and write the answer in the form (C, D). $\frac{C}{x-2} + \frac{D}{x+1} = \frac{6x}{x^2 - x - 2}$.

9. The rational expression $\frac{(x^2 + ax + b)}{(x^2 - 2ax + 2b)}$, where a and b are constants, reduces to $\frac{(x+3)}{(x+6)}$ with certain restrictions on x. Find the value of b.

Complex Numbers

> **For many centuries, zero didn't exist as a number.**

One of the amazing aspects of mathematics is that as you progressed in school, you continuously learned new concepts that were based on previous mathematics. For many centuries, zero didn't exist as a number. Imagine the problems that this caused in keeping monetary records. The first numbers that you learned as a young child dealt with counting objects. Then you learned fractions, negative numbers, and numbers like π and $\sqrt{2}$. You found that you could operate on all these different types of numbers. In algebra II, you learned a new number and you had thought that the reals was the last set. A mathematician wanted to solve the equation $x^2 = -1$. As we know, there is no solution in the real number system.

Problems in this unit deal with the **complex number** system. All real numbers are complex numbers, as you can write 5 as $5 + 0i$, etc. Listed below are the major ideas that you should review to solve the problems that follow.

1. The solution to the equation $x^2 = -1$ is $x = \sqrt{-1}$ and $\sqrt{-1} = i$, which is an imaginary number.
2. $\sqrt{-r} = \sqrt{r}i$, where r is a real number.
3. A number of the form $a + bi$ is a complex number; a and b are real numbers, and $i = \sqrt{-1}$. You can perform all four of the operations using complex numbers.
4. For addition and subtraction, you add/subtract the real number parts and then add/subtract the imaginary parts. For example, if $m = 2 - 6i$ and $n = 3 - 2i$, then
   $m + n = (2 - 6i) + (3 - 2i) = 5 - 8i$.
   $m - n = (2 - 6i) - (3 - 2i) = -1 - 4i$.
5. For multiplication, you multiply as you would $(2x - 3)(3x + 1)$, so
   $(2 - 6i)(3 - 2i) = 6 - 4i - 18i + 12i^2 = 6 - 22i - 12$, as
   $i^2 = -1$, the answer is $-6 - 22i$.

6. For division, $\frac{(2-6i)}{(3-2i)}$, you multiply both the numerator and denominator by the conjugate of the denominator, which is $3+2i$ as $3-2i$ is really $3-\sqrt{-4}$.
   Remember that $(bi)^2 = -b$ as $i^2 = -1$. Also,
   $(a+bi)(a-bi) = a^2+b^2$.

7. To simplify powers of i, use the concept that they are cyclic or repeat in cycles of 4 as
   $i = \sqrt{-1},\ i^2 = -1,\ i^3 = i^2 \cdot i = -1 \cdot i = -i,$
   $i^4 = i^2 \cdot i^2 = (-1)(-1) = 1,\ i^5 = i^4 \cdot i = i,$
   $i^6 = i^4 \cdot i^2 = -1$, etc.
   $\therefore\ i^{23} = (i^4)^5 \cdot i^3 = 1 \cdot i^3 = -i$. All you need to do is to divide the exponent by 4, and the remainder is the power of i, such as i^{98}, $\frac{98}{4} = 24$ with a remainder of 2, so
   $i^{98} = i^2 = -1$.

π TRY IT

Practice Activities

1. Evaluate $(1 - i)^{-8}$.

2. Solve for x in the domain of complex numbers:
   $5x + 2 = ix - 4i$.

3. If $Z = 3 + 2i$, determine the value of $Z^4 - Z^3 + Z^2 - Z + 1$.
   Express your answer in the form $a + bi$.

4. Find the square root of $15 - 8i$.

5. Solve: $ix^2 + 9x = 20i$.

6. Find the complex number Z for which the following equation is true: $Z + 6z = 7 + 3i$.
   Z is a complex number and z is its conjugate.

7. Find all ordered pairs (x, y) of real numbers that satisfy the equation
   $(x^2 - x - 5) + i(y - 12) = 1 - 7i$.

8. Evaluate: $(i^{(3^3)}) - (i^3)^3$.

Polynomial Functions

There are several major theorems listed below that deal with **polynomial functions.** Many of the concepts in this unit will be used in "Inverse Functions, Compositions, and Functions in General."

1. A polynomial function is of the form $p(x) =$ $a_nx^n + a_{n-1}x^{n-1} + a_{n-2}x^{n-2} + \ldots + a_2x^2 + a_1x + a_0$; a is a real number, n is a positive integer, and $a_n \neq 0$.

2. General rules for problems regarding polynomial functions:

   a) $p(r) = 0$ means r is a zero of p or r is a root of p.

   b) $x - r$ is a factor of p.

   c) $(r, 0)$ is an x-intercept on the graph of p.

There are several theorems that are important for problems dealing with polynomial functions.

There are several theorems that are extremely important for problems dealing with polynomial functions.

3. Remainder Theorem: For every polynomial $P(x)$ over the complex numbers, and for every complex number r, there exists a polynomial $Q(x)$ of one degree less, such that $P(x) = (x - r)Q(x) + P(r)$. $P(r)$ is a constant and is the remainder, and can be zero.

4. Factor Theorem: Over the set of complex numbers, $(x - r)$ is a factor.

5. The Fundamental Theorem of Algebra: For every polynomial with complex coefficients of degree n, there are exactly n complex roots.

6. The Rational Root Theorem: If p and q are integers and q is not equal to 0, and $\frac{p}{q}$ is reduced and is a rational root of a polynomial equation with integral coefficients, then p is a factor of the constant term or a_0 and q is a factor of the leading coefficient.

The majority of these theorems require the use of synthetic division.

For example,
$P(x) = x^3 - 2x^2 + 5x - 4$. Find all roots of $P(x)$. By the Rational Root Theorem, the possible roots are ± 1, ± 2, and ± 4. Always check ± 1 by inspection to see if they are roots, $P(1) = 1 - 2 + 5 - 4 = 0$, so 1 is a root; also, $x - 1$ is a factor of $P(x) = x^3 - 2x^2 + 5x - 4$. To find the other roots, it is usually easier to use synthetic division.

Using the Factor Theorem: Because 1 is a root, we divide the polynomial by 1.

1 |	1	–2	5	–4	List the coefficients; if one is missing, use 0.
		1			
	1				

Bring down the 1 and multiply 1 by the 1 (the root) and write it under the –2.

1 |	1	–2	5	–4	Add the –2 and 1 and place it under the 1.
		1	–1	4	Multiply $1(-1)$ and place it under the 5 and add.
	1	–1	4	0	Multiply $1(4)$ and place it under the –4.
					Add –4 and 4 and get 0.

$\therefore$ $P(1) = 0$ and, as we know, the remainder had to be 0. The 1, –1, 4 are the coefficients of the **depressed or reduced equation** and the degree of the equation is one less than the original.

$\therefore$ A new equation is formed: $x^2 - x + 4$; because this equation can't be factored, you can find the other roots by using the quadratic formula, $x = \dfrac{-b \pm \sqrt{b^2 - 4ac}}{2a}$ of a polynomial $P(x)$ if r is a root of $P(x) = 0$.

π TRY IT Practice Activities

1. Find the polynomial function $P(x)$ for which the zeros are ± 1, 3 and $P(0) = 9$.

2. If 2 is a root of $4x^3 - 3x^2 - kx - 4k^2 = 0$, find the values of k.

3. $P(x) = 3x^3 - 6x^2 - 3x + 6$. For which values of x is $P(x) > 0$?

4. Find $P(x)$ such that $P(1) = 6$ and $P(x) > 0$ only when $x < 0$ and $2 < x < 3$.

5. If $f(x) = x^4 + x^3 + ax + b$, find a and b such that $f(2) = -20$ and $f(-3) = 15$.

6. Given the polynomial function $f(x) = 2x^2 + ix - 1$, $i = \sqrt{-1}$, what is the remainder if $f(x)$ is divided by $x - i$?

7. Given the graph of a cubic function, $y = f(x)$ with zeros at 3 and 0 and $f(2) = 4$, find $f(4)$.

8. If $3x + 1$ is a factor of $6x^2 - kx - 1$, find k.

Inverse Functions, Compositions, and Functions in General

In this unit, you will solve problems involving algebraic **functions.** Many of the problems deal with inverses and composition of functions. You should review the definitions and theorems below and solve the problems.

1. **Composition of** two **functions,** f and g, is defined as $(f \circ g)(x) = f(g(x))$.

2. A function is said to be one to one if, for $x_1 \neq x_2$, then $f(x_1) \neq f(x_2)$.

3. If f is a one-to-one function, then a function f^{-1} exists and is called the **inverse function** of f. f^{-1} is the set of ordered pairs (y, x). $f^{-1}(y) = x$ if $y = f(x)$. The domain and range of the function f are switched in the function f^{-1}.

Example 1

If $F(x) = 2x + 3$, find $F^{-1}(x)$.

$F(x) = y = 2x + 3$; switch the x and y terms, then

$x = 2y + 3$, solve for y.

$2y = x - 3 \Rightarrow y = \frac{(x-3)}{2}$, $\therefore F^{-1}(x) = \frac{(x-3)}{2}$. Many students think that $F^{-1}(x)$ means the reciprocal of $F(x)$. $F^{-1}(x)$ is the inverse of $F(x)$.

4. If f and g are inverse functions, then $(f \circ g)(x) = x$ for all the numbers in the domain of g, and $(g \circ f)(x) = x$ for all numbers in the domain of f.

Example 2

Using $F(x) = 2x + 3$ and $F^{-1}(x) = \frac{(x-3)}{2}$; then

$F(F^{-1}(x)) = F\left(\frac{x-3}{2}\right) = 2\left(\frac{x-3}{2}\right) + 3 = x$ as the 2's cancel

and $x - 3 + 3 = 0$. Also,

$$F^{-1}(F(x)) = F^{-1}(2x+3) = \frac{2x+3-3}{2} = \frac{2x}{2} = x.$$

Example 3

If $f(3x) = 4x + \frac{3}{2}$, find $f\left(\frac{1}{2}\right)$.

You can't substitute $\frac{1}{2}$ for x, so do the following:

Let $y = 3x$, then $x = \frac{y}{3} \Rightarrow f(x) = 4\left(\frac{y}{3}\right) + \frac{3}{2} = \frac{4y}{3} + 3$,

$$\therefore f\left(\frac{1}{2}\right) = \frac{4 \cdot \frac{1}{2}}{3} + \frac{3}{2} = \frac{2}{3} + \frac{3}{2} = \frac{13}{6}.$$

π TRY IT Practice Activities

1. If $f(x) = 3x + 2$ and $g(x) = x^2 - 1$, find $g[f^{-1}(-2)]$.

2. If $f(x) = \dfrac{x(x-3)}{2}$, find the positive integer n such that $f(n) = 189$.

3. If $f(x) = x + 1$, $g(x) = 2x + 1$, and $h(x) = 3x + 1$, for what x does $3f(g(x)) + 2g(h(x)) = h(g(x))$?

4. If $f(x - 7) = 3x + 2a$, where a is a real number and $f(a) = 17$, find the numerical value of a.

5. If $f(n)$ is a function such that $f(1) = f(2) = f(3) = 2$ and $f(n+2) = \dfrac{f(n+1) \cdot f(n) + 1}{f(n-1)}$ for $n \geq 3$, find the numerical value of $f(6)$.

6. A function f has the following three properties for all positive integers for n: $f(1) = 1$, $f(2) = 3$, and $f(n) = 2f(n-1) - f(n-2)$ for $n \geq 3$. What is the value of $f(100)$?

7. If $p(x) = x^2 + 8$ and $q(x) = 2x$, for what value(s) of x is $p(q(x)) = q(p(x))$?

8. Given two functions f and g, $g(x) = 3x + 2$ and f is unknown. If $f(g(x)) = x^2 - x - 3$, what is the value of $f(1)$?

9. If $g(x) = -\frac{2}{3}(x + 5)$, then find $g^{-1}(-10)$.

Finding Zeros of Functions

If you have a polynomial function and you substitute a number in for the variables and get a value of 0, then that number is a zero of the function.

In addition to the information listed below, you should review the introduction for "Polynomials and Factoring." Many of these concepts and theorems will be used in this unit, especially the Remainder Theorem, the Factor Theorem, the Fundamental Theorem of Algebra, and the Rational Root Theorem. Synthetic division is used often.

If you have a polynomial function and you substitute a number in for the variables and get a value of 0, then that number is a root of the equation or a **zero of the function.**

Additional theorems:

1. If $P(x)$ represents a polynomial with odd degree, then $P(x) = 0$ has at least one real root.
2. For quadratic equations (degree 2):

   $ax^2 + bx + c = 0$, written in the form of $x^2 + \frac{b}{a}x + \frac{c}{a} = 0$,

   and with roots of r and s, then

   a) $r + s = -\frac{b}{a}$ or the sum of the two roots equals the middle term with its sign changed.

   b) $r \cdot s = \frac{c}{a}$ or the product of the two roots equals the constant term.

Example 1

Given $2x^2 + 3x - 5 = 0$, rewrite as $x^2 + \frac{3}{2}x - \frac{5}{2} = 0$, then the sum of the roots is equal to $-\frac{3}{2}$ and the product of the roots is equal to $-\frac{5}{2}$.

For equations greater than degree 2, the sum of the roots is the same, the quotient of the coefficients of the first two terms but the product of the roots is the same if the degree of the leading coefficient is even. If it is odd, then it is the opposite sign.

Example 2

Given $3x^3 - 2x^2 - x + 6 = 0$, the sum of the roots is $\frac{2}{3}$ and the product of the roots is $-\frac{6}{3} = -2$. However, for $x^4 + 3x^2 - 3x - 4 = 0$, the sum of the roots is 0 because there is no cubic term (or 3rd degree) and the product of the roots is –4. Remember that, if $a + bi$ is a root, then its conjugate $a - bi$ is also a root. Also, if $a + \sqrt{b}$ is a root, then $a - \sqrt{b}$ is a root.

For quadratics, it is helpful to remember that
$x^2 -$ (sum of the roots)x + (product of the roots) $= 0$.

The most efficient way to find the quadratic equation is to use the rule for sum and product of the roots.

Example 3

Find the quadratic equation with $3 \pm 2i$ as roots.

The most efficient way to find the quadratic equation is to use the rule for sum and product of the roots. Let $r = 3 + 2i$ and $s = 3 - 2i$. The $r + s = (3 + 2i) + (3 - 2i) = 6$, so the middle term is –6, and $r \cdot s = (3 - 2i)(3 + 2i) = 9 - 4i^2 = 13$, so the equation is $x^2 - 6x + 13 = 0$.

Given the quadratic formula, $x = \frac{-b \pm \sqrt{b^2 - 4ac}}{2a}$, $b^2 - 4ac$ is called the **discriminant** and is referred to as D. D controls the number and type of the root.

1. If $D = 0$, there is a **double root.**
2. If $D < 0$, there are two complex conjugate roots.
3. If $D > 0$, there are two different real roots.

Also, if the quadratic equation has rational coefficients and D is a perfect square, then the roots are rational.

Example 4

1. Find the type and number of roots for $2x^2 - 3x + 5 = 0$.
   $b^2 - 4ac = (-3)^2 - 4(2)(5) = 9 - 40 = -31$. Therefore, $2x^2 - 3x + 5 = 0$ has two complex conjugate roots.

2. Change the equation to $2x^2 - 3x - 5 = 0$, then
   $b^2 - 4ac = (-3)^2 - 4(2)(-5) = 9 + 40 = 49$; because the coefficients are rational and $D > 0$ and is a perfect square, there are two rational roots.

Practice Activities

1. The product of two of the roots of the equation $x^3 + 2x + 12 = 0$ is 6. Find the roots.

2. The sum of the squares of the roots of the equation $x^2 + 4hx = 5$ is 154. Compute $|h|$.

3. If the roots of $6x^2 - 13x + 6c = 0$ are r and $\frac{1}{r}$, find the smaller such r.

4. One root of $4x^3 - 8x^2 + cx + d = 0$ is -1. The other two roots are equal. Find d.

5. Let $f(x) = \frac{(x+4)^2}{x-3}$ and $g(x) = \frac{2x^2 - 12x - 31}{x-3}$; if $h(x) = f(x) + g(x)$, find the zeros of $h(x)$.

6. For what value(s) of k does the equation $2x^2 - 2kx + 5k = 0$ have one root?

7. Find all positive integral values of B for which the following quadratic equation has two distinct integral roots: $x = \frac{B}{(10 - x)}$.

8. $4 - \sqrt{3}$ is one root of $2x^2 + bx + c = 0$, where b and c are rational numbers. Find the value of $b + c$.

Exponential Functions

If you have an exponential equation, change the bases to the same base and then set the exponents equal to each other.

Review all the rules for exponents in "Polynomials and Factoring." The definition of an **exponential function** is $y = a^x$, $a \neq 1$ and $a > 0$, and is a real number. Its domain (x's) is the reals and its range (y's) are positive numbers.

The problems below involve concepts with exponentials and allow you to solve problems involving 2^π, $8^{\sqrt{2}}$ and so on.

In this unit, we will make use of the following laws of exponents:

1. If r is a positive integer and b a positive real number, then $\sqrt[r]{b} = b^{\frac{1}{r}}$.
2. If the above conditions exist and p is an integer, then $b^{\frac{p}{r}} = \sqrt[r]{b^p} = (\sqrt[r]{b})^p$; remember the placements of the (p)ower and (r)oot.
3. For $b > 0$ and $b \neq 1$, if $b^x = b^y$, then $x = y$. Use the above laws and those from "Exponents and Radicals" to solve the following problems.

A few hints:

Change the bases in the problems to the lowest bases; 8, 32, 16 can be expressed in base 2. If you have an **exponential equation,** change the bases to the same base and then set the exponents equal to each other. Be careful solving this unit's problems because it is very easy to make errors.

Example 1

Express in simplest radical form:

$$\sqrt[4]{27} \cdot \sqrt[3]{9} = 27^{\frac{1}{4}} \cdot 9^{\frac{1}{3}} = (3^3)^{\frac{1}{4}}(3^2)^{\frac{1}{3}} = 3^{\frac{3}{4}} \cdot 3^{\frac{2}{3}} = 3^{\frac{3}{4}+\frac{2}{3}} =$$
$$3^{\frac{9}{12}+\frac{8}{12}} = 3^{\frac{17}{12}} = \sqrt[12]{3^{17}} = 3\sqrt[12]{3^5} = 3\sqrt[12]{243}.$$

Example 2

Solve for x: $4^{2x-3} = \left(\frac{1}{4}\right)^{x+1}$.

$(2^2)^{2x-3} = (2^{-2})^{x+1}$, $2^{4x-6} = 2^{-2x-2} \Rightarrow 4x - 6 = -2x - 2$.

$6x = 4$, $\therefore x = \frac{2}{3}$.

Example 3

Solve for x.

$x^{\frac{2}{3}} - x^{\frac{1}{3}} - 6 = 0$.

Solution 1:

You can let $a = x^{\frac{1}{3}}$ and get $a^2 - a - 6 = 0$; factor

$(a-3)(a+2) = 0$, $\therefore a = -2, 3$; because $a = x^{\frac{1}{3}}$, then $x^{\frac{1}{3}} = 3$;

$\left(x^{\frac{1}{3}}\right)^3 = 3^3 \Rightarrow x = 27$, and $x^{\frac{1}{3}} = -2$. This is rejected as $(-8)^{\frac{2}{3}}$ is not defined. $\therefore x = 27$ is the only answer.

Solution 2:

Factor $x^{\frac{2}{3}} - x - 6 = 0$, $\left(x^{\frac{1}{3}} - 3\right)\left(x^{\frac{1}{3}} + 2\right) = 0$; then proceed as above.

π TRY IT

Practice Activities

1. Solve for m: $3^{m+2} = 3^m + 1{,}944$.

2. Find x: $2^{x+y} = 256$ and $3^{x-y} = 729$.

3. Solve for x: $\left(\frac{1}{16}\right)^{x-3} = (32)^{x+3}$.

4. Solve for x: $\sqrt{125^x} = \frac{5}{25^x}$.

5. Find all real numbers x for which $\frac{3^{\sqrt{12x}} + 3}{4} = 3^{\sqrt{3x}}$.

6. Solve for x: $\sqrt{\frac{9^{x+3}}{27^x}} = 81$.

7. Given $5^x + 2^y = 2^x + 5^y = \frac{7}{10}$; find $(x + y)^{-1}$.

8. Find x: $4^{x+1} - 5 \cdot 2^x + 1 = 0$.

Logarithmic Functions

It is common to write logs in base 10 without showing the 10 in the base.

The logarithmic function with base b is the inverse function of the exponential function with base b. If $x = b^y$, then the **logarithm** to the base b of x is y and is written $y = \log_b x$, and both b and x must be greater than 0. This is extremely important because, when solving log problems, you obtain negative numbers for b and x, which are extraneous answers and must be rejected. Remember that logs are exponents. It is common to write logs in base 10 without showing the 10 in the base; $\log 100 = 2$ because $10^2 = 100$.

Laws of logarithms:

For all the laws below, $b \neq 1$, $b > 0$.

1. $\log_b a = x$ is equivalent to $b^x = a$.
2. $\log_b b^x = x$ $(b > 0, b \neq 1)$ and $b^{\log_b x} = x$ (same and $x > 0$). An easy way to remember this law is to notice where the b's are located.
3. $\log_b(A \cdot B) = \log_b A + \log_b B$. The log of a product is the sum of their logs.
4. $\log_b \frac{A}{B} = \log_b A - \log_b B$. The log of a quotient is the difference of their logs.
5. $\log_b A^n = n \cdot \log_b A$.
6. $\log_b A^x = \log_b A^y$ means that $x = y$ and $x, y > 0$. Equality of logs.
7. $\log_b A = \frac{(\log_c A)}{(\log_c b)}$. This is the law for changing the base of a log to another base. You can change to any base as long it is a positive base. It is easy to remember, since b is the bottom and a is on top.

There are other laws, but these are the ones used most often.

π TRY IT Practice Activities

1. Find all real x's such that $\log_3(x+3) = 1 - \log_3(x+5)$.

2. Solve for x: $\log_2[\log_3(x+1)] = -1$.

3. What is the simplified numerical value of the sum

   $\log_2\left(1-\frac{1}{2}\right) + \log_2\left(1-\frac{1}{3}\right) + \ldots\log_2\left(1-\frac{1}{64}\right)$?

4. If $\log_{10}2 = a$ and $\log_{10}3 = b$, find the exact solution to the equation $12^{x+2} = 18^{x-3}$ in terms of a and b.

5. If $\log_{10}x^n = 100x^2$ and $\log_x 10 = n$, find x.

6. Solve for x: $x^{\log_2 x} = 16x^3$.

7. Solve: $\dfrac{625}{x^{\log_5 x}} = 1$.

8. Let $r = \log_b\left(\frac{8}{45}\right)$ and $s = \log_b\left(\frac{135}{4}\right)$. Find an ordered pair of integers (m, n) such that

   $\log_b\left(\frac{32}{5}\right) = mr + ns$.

9. If $(\log_3 x)(\log_x 2x)(\log_{2x} y) = \log_x x^2$, compute the numerical value of y.

10. Given $\log_{18}6 = a$, find $\log_{18}16$ in terms of a.

Solving Systems of Equations

This unit is an extension of the traditional solving of systems of equations that you did in previous courses. You will use the same methods, elimination or substitution. This unit offers a wide variety in solving different types of systems—not simply linear-linear, but linear-quadratic and problems of a different nature.

Example 1

Find the four real solutions (x, y) for the following system of equations: $x + xy + y = 11$ and $x^2y + xy^2 = 30$.

Solution:

Notice that if you factor the second equation, $x^2y + xy^2 = 30$, $xy(x + y) = 30$, you can let $a = xy$ and $b = x + y$, then $a + b = 11$ and $ab = 30$; solve the first equation for a, $a = 11 - b$; and substitute in the second equation, $(11 - b)b = 30$, $b^2 - 11b + 30 = 0$; factor, $(b - 6)(b - 5) = 0 \Rightarrow b = 5, 6$. $\therefore$ Because $ab = 30$, $a = 6, 5$. Then $x + y = 5$ and $xy = 6 \Rightarrow (2, 3), (3, 2)$, and $x + y = 6$ and $xy = 5 \Rightarrow (5, 1), (1, 5)$. $\therefore$ The four solutions are $(2, 3), (3, 2), (5, 1), (1, 5)$.

Example 2

Find the ordered triple of real numbers (x, y, z) that satisfies $x^2y^2z = 225$, $xy^2z^2 = 75$, and $x^2yz^2 = 45$.

Solution:

Multiply the three equations together. $x^5y^5z^5 = 225 \cdot 75 \cdot 45$; factor into primes, $225 \cdot 75 \cdot 45 = (3 \cdot 5)^2 \cdot 5^2 \cdot 3 \cdot 3^2 \cdot 5 = 5^5 \cdot 3^5$. $\therefore x^5y^5z^5 = 5^5 \cdot 3^5$; take the fifth root of both sides, and get $xyz = 5 \cdot 3$. $\therefore z = 1$. Divide $x^2y^2z = 225$ by $xy^2z^2 = 75 \Rightarrow \frac{(x^2y^2z)}{(xy^2z^2)} = \frac{225}{75}$ and get $x = 3$; also, $\frac{x^2y^2z}{x^2yz^2} = \frac{225}{45} \Rightarrow y = 5$. $\therefore (x, y, z) = (3, 5, 1)$.

π TRY IT Practice Activities

1. Solve for the ordered pair (x, y).

   $\frac{6}{x} + \frac{5}{y} = 1$

   $\frac{3}{x} - \frac{10}{y} = 3$

2. Find all ordered pairs (t, w) that make the following statements true:

   $\frac{2}{t} = 7 - \frac{3}{2w}$ and $\frac{6}{w} + \frac{1}{t} = 0$.

3. Solve for the ordered pair (x, y).

   $2x + \sqrt{5}y = 7$ and $\sqrt{5}x - 3y = -2\sqrt{5}$.

4. Find all ordered pairs (x, y), in terms of a, that satisfy the following system of equations:

   $3ax + 4ay = 1$ and $\frac{x}{3} - \frac{y}{2} = \frac{2}{a}$.

5. The graphs of $y = -x^2 + x - 1$ and $\frac{(y+3)}{(x-2)} = 2$ have only one point of intersection. Find this ordered pair, (x, y).

6. Find the value of $x^2 + y^2$ if (x, y) is a solution of the system of equations: $xy = 5$ and $x^2y + 2x = xy^2 + 2y + 35$.

7. Find the intersection points (in simplest radical form) of the circle $x^2 + y^2 = 40$ and $x^2 + 2xy - 3y^2 = 0$.

8. If $abc > 0$, find the ordered triple that satisfies $ab = 24$, $ac = 72$, and $bc = 108$.

Matrices and Determinants

If two matrices have the same dimension, we can add and subtract them.

All the problems in this unit can be solved by using a graphing calculator. You should know how to solve them using a calculator, but the problems are written to solve without using a calculator.

A rectangular array of numbers expressed as $C = \begin{bmatrix} 1 & \sqrt{2} \\ 4 & \pi \\ 0 & -6 \end{bmatrix}$ is a **matrix.** Its **dimension** is determined by its number of rows and its number of columns. Matrix C has dimension $C_{3\times 2}$.

Operations:

If two matrices have the same dimension, we can add and subtract them.

Example 1

$$A = \begin{bmatrix} 3 & 4 & 1 \\ -7 & -3 & 0 \end{bmatrix}, B = \begin{bmatrix} 4 & -8 & 12 \\ 6 & -5 & -4 \end{bmatrix}.$$

$$\therefore\ A + B = \begin{bmatrix} 7 & -4 & 13 \\ -1 & -8 & -4 \end{bmatrix} \text{ and } A - B = \begin{bmatrix} -1 & 12 & -11 \\ -13 & 2 & 4 \end{bmatrix}.$$

A matrix can be multiplied by a real number, which is called a scalar.

$$3\begin{bmatrix} 7 & -4 & 13 \\ -1 & -8 & -4 \end{bmatrix} = \begin{bmatrix} 21 & -12 & 39 \\ -3 & -24 & -12 \end{bmatrix}.$$

In order to multiply matrices, the following rule involving their dimensions must exist:

The product of a matrix $A_{a\times b}$ and matrix $B_{b\times c}$ is matrix $C_{a\times c}$. The column dimension of A must be the same as the row dimension of B, and the answer matrix C will have dimension of the row of A and the column of B.

$\begin{bmatrix} a & b \\ c & d \end{bmatrix} \times \begin{bmatrix} u & v \\ w & x \end{bmatrix} = \begin{bmatrix} au + bw & av + bx \\ cu + dw & cv + dx \end{bmatrix}$; remember to multiply row by column.

Example 2

Multiply $\begin{bmatrix} 2 & 4 & -1 \\ 0 & -3 & 5 \end{bmatrix} \times \begin{bmatrix} 2 \\ 3 \\ 4 \end{bmatrix}$.

Call the first matrix A, the second B, and the answer C.
$\therefore A_{2\times 3} \times B_{3\times 1} = C_{2\times 1}$.

$$\begin{bmatrix} 2 & 4 & -1 \\ 0 & -3 & 5 \end{bmatrix} \times \begin{bmatrix} 2 \\ 3 \\ 4 \end{bmatrix} = \begin{bmatrix} 2 \cdot 2 + 4 \cdot 3 - 1 \cdot 4 \\ 0 \cdot 2 - 3 \cdot 3 + 5 \cdot 4 \end{bmatrix} = \begin{bmatrix} 4 + 12 - 4 \\ 0 - 9 + 20 \end{bmatrix} = \begin{bmatrix} 12 \\ 11 \end{bmatrix}.$$

$A = \begin{bmatrix} a & b \\ c & d \end{bmatrix}$, then $\begin{vmatrix} a & b \\ c & d \end{vmatrix}$ is the **determinant** of matrix A and equals the number obtained from $ad - cb$.

The determinant of $\begin{vmatrix} a & b & c \\ d & e & f \\ g & h & i \end{vmatrix}$ is found by repeating the first two columns and using the above rule.

$$\begin{matrix} a & b & c & a & b \\ d & e & f & d & e \\ g & h & i & g & h \end{matrix} = aei + bfg + cdh - (gec + hfa + idb).$$

There are many applications using matrices and determinants. They are very useful in solving systems of equations.

π TRY IT Practice Activities

1. Find the value of x: $\begin{vmatrix} 2 & 3 \\ 7 & 1 \end{vmatrix} = \begin{vmatrix} 4 & 6 \\ 14 & x \end{vmatrix}$.

2. Find all values of x for which $\dfrac{\begin{vmatrix} x & 2 \\ 2 & x \end{vmatrix}}{\begin{vmatrix} x & x \\ x & 2 \end{vmatrix}} = -\dfrac{5}{3}$.

3. Solve for x and y. Express your answer as an ordered pair (x, y).
   $\begin{vmatrix} x-3 & y+1 \\ 17 & -3 \end{vmatrix} = -3$ and $\begin{vmatrix} y-2 & x+4 \\ 2 & -19 \end{vmatrix} = 18$.

4. Evaluate the determinant: $\begin{vmatrix} 0 & 2 & -3 \\ 3 & 5 & -3 \\ 1 & 2 & 0 \end{vmatrix}$.

5. Solve for x: $\begin{vmatrix} x^2 & 3x & 5 \\ 1 & 3 & 5 \\ 4 & -6 & 5 \end{vmatrix} = 0$.

6. If $A = \begin{bmatrix} 1 & 3 \\ 4 & 3 \end{bmatrix}$ and $B = \begin{bmatrix} x+1 \\ y \end{bmatrix}$, find x and y such that $A \cdot B = 3 \cdot B$.

7. Compute: $3\begin{bmatrix} 2 & -5 & 1 \\ 3 & 0 & -4 \end{bmatrix} - 2\begin{bmatrix} 1 & -2 & -3 \\ 1 & -1 & 4 \end{bmatrix}$.

8. Find the ordered pair (x, y) such that $\begin{bmatrix} x & y \\ 2 & 4 \end{bmatrix} \cdot \begin{bmatrix} 1 & y-4 \\ 2 & x \end{bmatrix} = \begin{bmatrix} -1 & -50 \\ 10 & 6 \end{bmatrix}$.

Sequences and Series

In this unit, you will solve problems dealing with **arithmetic and geometric sequences and series.** This unit involves many definitions and formulas that are listed below. Suggestions are given as to their use.

An arithmetic sequence or progression is a sequence in which the difference of consecutive terms is the same.

1. Summation notation: capital sigma of the greek alphabet,
   $$\sum_{n=1}^{k} m_n = m_1 + m_2 + m_3 + \ldots + m_k.$$

2a. An arithmetic sequence or progression is a sequence in which the difference of consecutive terms is the same. This number is called the common difference and is denoted by d. Examples are $a_1,\ a_1 + d,\ a_1 + 2d,\ \ldots,\ a_1 + (n-1)d$ or 1, 3, 5, 7.

2b. To find the nth term (a_n), first term (a_1), number of terms (n), or the common difference (d), you can use $a_n = a_1 + (n-1)d$.

2c. The terms between any two terms are called the **arithmetic means;** example: 3 is the arithmetic mean between 1 and 5, and 3 and 5 are the arithmetic means between 1 and 7. The familiar average of two or more scores is a use of the arithmetic mean.

3a. An **arithmetic series** is the sum of an arithmetic sequence and is denoted by S_n; $S_4 = 16$ for $1 + 3 + 5 + 7$.

3b. To find the sum of a **finite sequence,** use
   $$S_n = \frac{n}{2}[2a_1 + (n-1)d] \text{ or } S_n = \frac{n}{2}(a_1 + a_n).$$

4a. A geometric sequence or progression is a sequence in which the quotient of each term divided by the preceding one is the same constant. This constant is called the common ratio and is denoted by r. Examples are $a_1,\ a_1r^1,\ a_1r^2,\ \ldots,\ a_1r^{n-1}$ or $2,\ 2^2,\ 2^3,\ 2^4 = 2, 4, 8, 16$.

4b. To find the common ratio (r), or a particular term, use $a_n = a_1r^{n-1}$.

4c. The sum of an infinite **geometric series,** $|r| < 1$, is $S = \frac{a_1}{1-r}$.

Examples: Do these before you do the problem set. Use the formulas if needed. The solutions follow the problems.

Example 1

What is the first term of an arithmetic sequence for which the fifth term is 14 and the eighth term is 23?

Solution:

Use the formula $a_n = a_1 + (n-1)d$. You don't know d or the first term, but you can create a system of equations by using 14 and 23 as a_n.

a. $14 = a_1 + (5-1)d \Rightarrow 14 = a_1 + 4d$

b. $23 = a_1 + (8-1)d \Rightarrow 23 = a_1 + 7d$; subtract equations *a* and *b*. $9 = 3d \Rightarrow d = 3$. Use equation *a* to find a_1. $\therefore a_1 = 2$.

Example 2

What is the sixth term in the geometric sequence 3, 12, 48, ... ?

Solution:

Use the formula, don't just multiply by 4 until you get to the sixth term.

Common ratio $r = \frac{12}{3} = 4$, $a_1 = 3$, and $n = 6$.

$(\therefore a_n = a_1 r^{n-1}) \Rightarrow a_6 = 3 \cdot 4^5 = 3 \cdot 1{,}024 = 3{,}072$.

Example 3

In the geometric sequence $\frac{1}{x}, \frac{1}{(4x+4)}, \frac{1}{(8x+8)}$ ..., what is the common ratio?

Solution:

You can form a proportion by using $\dfrac{\frac{1}{8x+8}}{\frac{1}{4x+4}} = \dfrac{\frac{1}{4x+4}}{\frac{1}{x}}$, common ratios are equal, and simplifying, $\dfrac{4x+4}{8x+8} = \dfrac{x}{4x+4}$; solving gives $x = -2$, so $r = \dfrac{1}{2}$.

Example 4

Find the values of x, $x \neq 0$, for which $x^2 - x^3 + x^4 - x^5 \ldots = \dfrac{x}{5}$.

Solution:

This is an infinite geometric series, so use $S = \dfrac{a_1}{(1-r)}$.

$\therefore \dfrac{x}{5} = \dfrac{x^2}{1-(-x)}, \dfrac{x}{5} = \dfrac{x^2}{1+x}$; cross multiply, $x + x^2 = 5x^2$, $x = \dfrac{1}{4}$ as $x \neq 0$.

π TRY IT Practice Activities

1. If $\frac{1}{3}$, $\frac{1}{x}$, and $\frac{1}{7}$ form an arithmetic progression in that order, compute x.

2. Find $\sum_{n=1}^{12}\left(\frac{1}{n+2}-\frac{1}{n+3}\right)$.

3. The smallest interior angle of a polygon with 54 diagonals measures 132°. If the measures of all the interior angles are in arithmetic progression (A.P.), find the measure of the largest angle of the polygon.

4. Find the first term of an infinite geometric progression for which the sum is $2\sqrt{2}+2$ and for which the common ratio is $\frac{1}{\sqrt{2}}$.

5. Find all possible values of X and Y such that 3, X, and Y will be in an arithmetic progression and X, Y, and 8 will be in a geometric progression. Express the answer(s) as ordered pair(s) (X, Y).

6. The sum of n positive terms of an arithmetic series is 216. The first term is n and the last term is twice the first term. What is the common difference?

7. An infinite geometric progression has a first term 1, and the sum of all terms is $\frac{9}{2}$ times its second term. What possible values could the common ratio have?

8. In a geometric series of positive terms, the difference between the fifth and fourth term is 243, and the difference between the second and first term is 9. What is the sum of the first five terms of the series?

9. The arithmetic mean of two positive numbers exceeds their geometric mean by 50. By how much does the square root of the larger exceed the square root of the smaller?

Analytical Geometry

Analytical geometry covers a wide area of mathematics.

Analytical geometry covers a wide area of mathematics. The problem section contains problems dealing with linear equations, circles, parabolas, ellipses, and hyperbolas. Information contained here will only pertain to the problems that you are asked to solve.

Listed are important definitions and theorems for this unit:

1. The slope-intercept formula for finding the equation of a line is $y = mx + b$, and the intercept form of a linear equation is $\frac{x}{a} + \frac{y}{b} = 1$, where a and b are the x- and y-intercepts, respectively.

2. If (x_1, y_1) and (x_2, y_2) are points on the coordinate axes, then the distance between them is $D = \sqrt{(x_1 - x_2)^2 + (y_1 - y_2)^2}$.

3. Distance formula from a point (x, y) to a line $(Ax + By = C)$ is $D = \frac{|Ax + By + C|}{\sqrt{A^2 + B^2}}$; (x, y) is the ordered pair; and A, B, and C are the coefficients of the linear equation. Make sure that you rewrite the linear equation as $Ax + By + C = 0$.

Circles:

4. $x^2 + y^2 = r^2$ is the equation of a circle, the center of which is at the origin and the radius of which is r.

5. $(x - h)^2 + (y - k)^2 = r^2$ is the equation for the circle with center at the ordered pair (h, k) and with the radius r.

6. The general form of a circle is $x^2 + y^2 + Ax + By + C = 0$. In order to find the center and radius of the equation, we use a procedure called completing the square.

Example 1

Find the center and radius of $x^2 + y^2 + 2x - 6y + 6 = 0$.
$x^2 + y^2 + 2x - 6y + 6 = 0$; rewrite as
$x^2 + 2x + \_\_\_ + y^2 - 6y + \_\_\_ = -6$; complete the square of each polynomial and add the two constants to the –6.

$x^2 + 2x +$ ___ $+ y^2 - 6y +$ ___ $= -6$; take $\frac{1}{2}$ of the middle terms; square the answer; and add to each polynomial.
$x^2 + 2x + 1 + y^2 - 6y + 9 = -6 + 1 + 9$; factor the polynomials; and combine the constants. $\therefore\ (x+1)^2 + (y-3)^2 = 4$. The center is $(-1, 3)$ and the radius is 2.

Parabolas:

7. The general equation of the parabola or quadratic function is $y = ax^2 + bx + c$. When we complete the square, we get $y = a\left(x + \frac{b}{2a}\right)^2 + c - \frac{b^2}{4a}$. The vertex is $\left(-\frac{b^{\ 2}}{2a}, c - \frac{b^2}{4a}\right)$.

   An easier formula is $\left(-\frac{b}{2a}, f\left(-\frac{b}{2a}\right)\right)$, where you compute $-\frac{b}{2a}$, then substitute its value for x to find the y coordinate.

8. Another form of a parabola is $y = a(x-h)^2 + k$, where (h, k) is the vertex, $x = h$ is the axis of symmetry, and k is the minimum or maximum value.

Ellipses:

9. The general form is $Ax^2 + By^2 + Cx + Dy + E = 0$. There are four basic formulas depending on whether the x-axis or y-axis is the major axis and whether the ellipse is centered at the origin or at some point (x, y).

Ellipses with the center at the origin:

10a. $\frac{x^2}{a^2} + \frac{y^2}{b^2} = 1$; ellipses must be in this form, $^*a > b$ in all the ellipse formulas.

The major axis is the x-axis and has length $2a$. The vertices (x-intercepts) are $(a, 0)$ and $(-a, 0)$. The minor axis is the y-axis and has length $2b$.

The y-intercepts are $(0, b)$ and $(0, -b)$.

There are two ordered pairs $(c, 0)$ and $(-c, 0)$ called foci. You can determine c by using the formula: $c = \sqrt{a^2 - b^2}$.

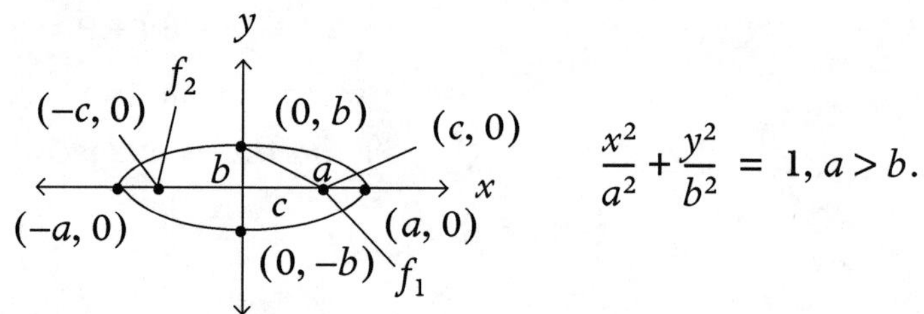

10b. $\dfrac{x^2}{b^2} + \dfrac{y^2}{a^2} = 1, a > b.$

The major axis is the y-axis and still has length $2a$, so the vertices are $(0, a)$ and $(0, -a)$.
The minor axis is the x-axis and still has length $2b$ so the x-intercepts are $(b, 0)$ and $(-b, 0)$.
The formula to find the focus point is the same.

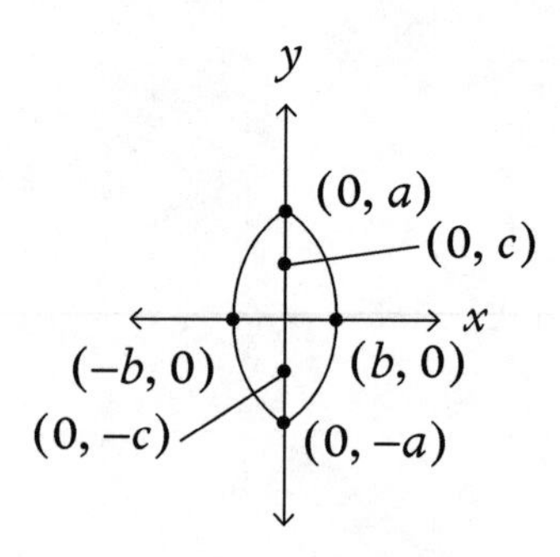

11a. $\dfrac{(x-h)^2}{a^2} + \dfrac{(y-k)^2}{b^2} = 1, a > b$. The major axis is parallel to the x-axis. The center is at (h, k) and the information given in 10 is still applicable.

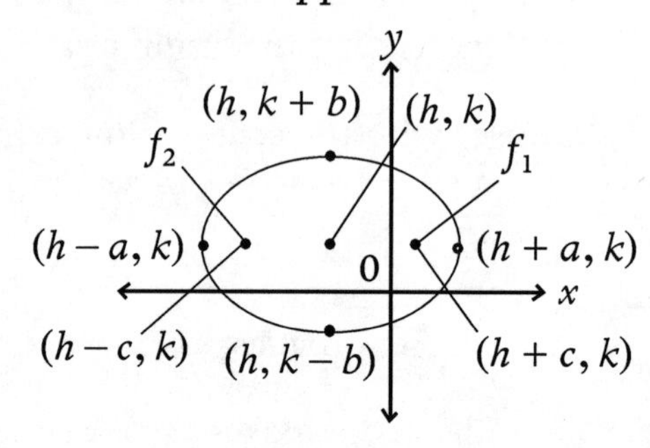

11b. $\frac{(x-h)^2}{b^2} + \frac{(y-k)^2}{a^2} = 1, a > b$. The major axis is parallel to the y-axis, the center is at (h, k), and information in 10 holds.

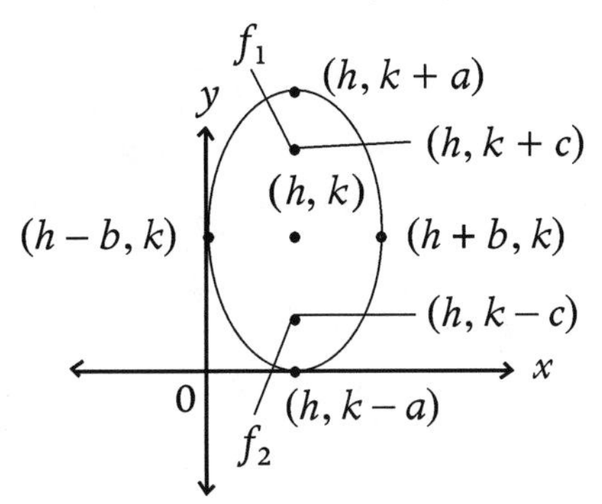

Example 2

Given $\frac{(x-2)^2}{25} + \frac{(y-3)^2}{9} = 1$, find the center, vertices, endpoints of the minor axis, and foci, and sketch the ellipse.

The center is $(2, 3)$.

The major axis has length $2a$, $a = 5$, so $2a = 10$. The vertices are $(h \pm a, k)$, so they are $(7, 3)$ and $(-3, 3)$.

The minor axis has length $2b$, $b = 3$, so $2b = 6$.

The end points for the minor axis are $(h, k \pm b)$, so they are $(2, 6)$ and $(2, 0)$.

Focus: The ordered pairs for the foci are $(h \pm c, k)$ and $c = \sqrt{a^2 - b^2}$. $\therefore$ $c = \sqrt{25 - 9} = \pm 4$. $\therefore$ The focus points are at $(6, 3)$ and $(-2, 3)$. You can remember all this information if you sketch the graph first, locate the endpoints of the major and minor axes, and then generate the formula to find the focus of a and b. See the diagram by using length,

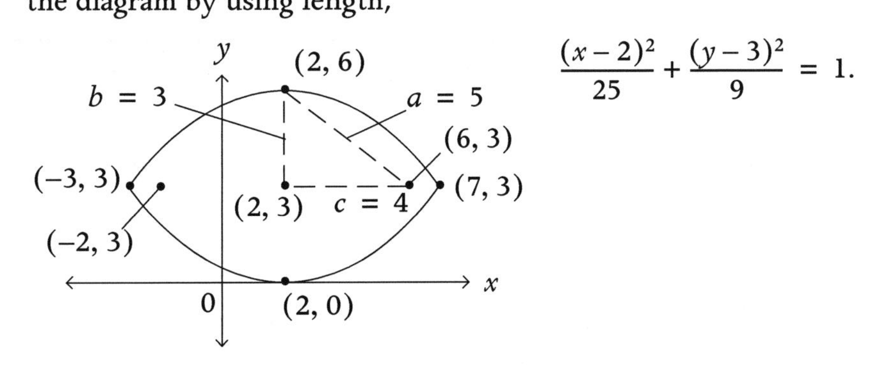

$$\frac{(x-2)^2}{25} + \frac{(y-3)^2}{9} = 1.$$

Hyperbolas:

The general form of a hyperbola is
$Ax^2 + Bxy + Cy^2 + Dx + Ey + F = 0$.

Similar to ellipses, there are four variations of hyperbolas, whether they are centered about the origin or about an ordered pair other than the origin.

Hyperbolas with center at the origin:

12a. $\frac{x^2}{a^2} - \frac{y^2}{b^2} = 1$; hyperbolas must be in this form.

The x-axis is called the transverse axis, with length $2a$, and has vertices of $(a, 0)$ and $(-a, 0)$.
The y-axis is the conjugate axis (think of complex numbers) and has no y-intercepts.

The asymptotes are $y = \pm\left(\frac{b}{a}\right)x$.

The focus points are $(c, 0)$ and $(-c, 0)$, and the formula to find c is the Pythagorean Theorem $c^2 = a^2 + b^2$ or
$c = \sqrt{a^2 + b^2}$.

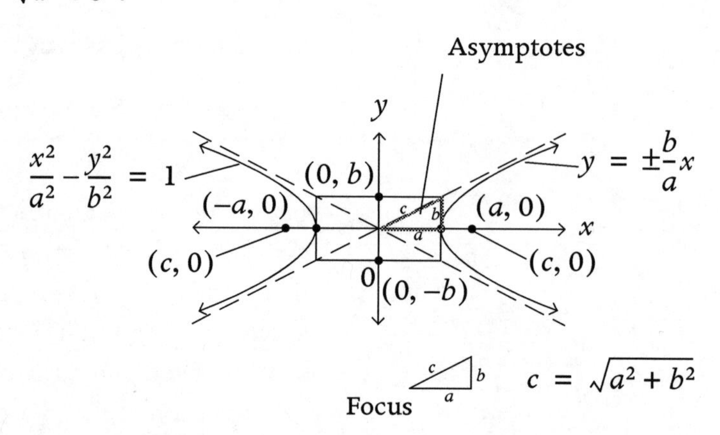

12b. $\frac{y^2}{a^2} - \frac{x^2}{b^2} = 1$

The y-axis is the transverse axis, with length $2a$, and has vertices $(0, a)$ and $(0, -a)$.
The x-axis is the conjugate axis and has no x-intercepts.

The asymptotes are $y = \pm\left(\frac{a}{b}\right)x$.

The focus points are $(0, c)$ and $(0, -c)$ and the formula is the same.

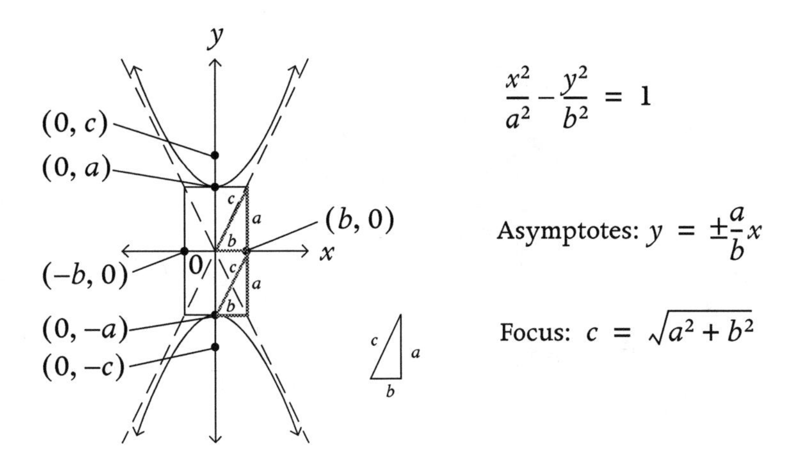

Hyperbolas also have centers other than the origin. There are two types with center at (h, k) and the tranverse axis is parallel to either the x-axis or y-axis. The two equations are:

13a. $\frac{(x-h)^2}{a^2} - \frac{(y-k)^2}{b^2} = 1$, center (h, k), vertices $(h \pm a, k)$,

and asymptotes the lines $y = \pm\left(\frac{b}{a}\right)(x-h) + k$.

$$\frac{(x-h)^2}{a^2} + \frac{(y-k)^2}{b^2} = 1, a > b$$

y

$y = \frac{b}{a}(x-h) + k$

$(h - a, k)$ $(h + a, k)$

x

(h, k)

$y = -\frac{b}{a}(x-h) + k$

13b. $\dfrac{(y-k)^2}{a^2} - \dfrac{(x-h)^2}{b^2} = 1$, center (h, k), vertices $(h, k \pm a)$, and asymptotes the lines $y = \pm\left(\dfrac{a}{b}\right)(x-h) + k$.

$$\frac{(y-k)^2}{a^2} - \frac{(x-h)^2}{b^2} = 1$$

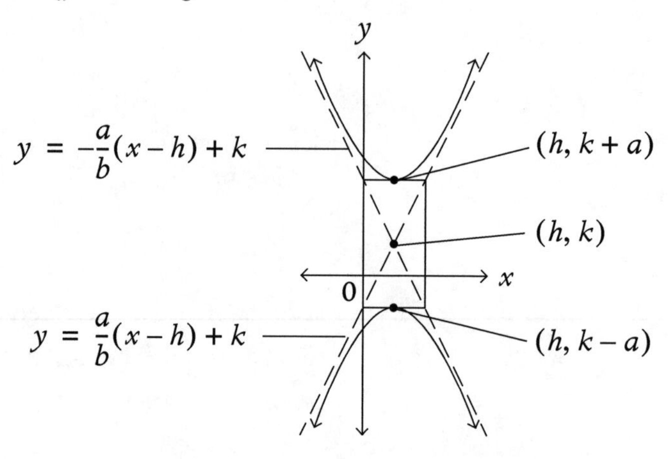

Example 3

Sketch $\dfrac{(x-2)^2}{9} - \dfrac{(y-1)^2}{4} = 1$.

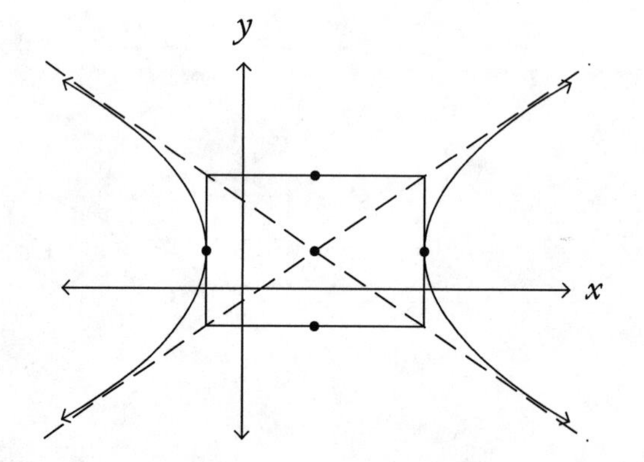

Center $(2, 1)$.

Vertices $(5, 1)$, $(-1, 1)$.

Asymptotes:
$y = \pm\frac{2}{3}(x-2) + 1$.

Example 4

$\frac{x^2}{9} - \frac{y^2}{16} = 1$. Find the vertices, foci, asymptotes, and sketch the hyperbola.

Vertices are $(\pm 3, 0)$.

Use $c = \sqrt{a^2 + b^2}$ to find the foci; $c = \sqrt{9 + 16} = \pm 5$, so foci are $(\pm 5, 0)$.

Asymptotes are $y = \pm\left(\frac{4}{3}\right)x$. You can memorize the formula or solve for y:

$\frac{x^2}{9} = \frac{y^2}{16}$; notice that we didn't use the 1, $9y^2 = 16x^2$, $y^2 = \frac{16}{9}x^2$,

and $y = \pm\left(\frac{4}{3}\right)x$ (this works for all ellipses and also for those where the center is (x, y)).

You can determine the asymptotes for which the center is the origin by using the sketch. Locate (a, b) and use $\frac{\text{rise}}{\text{run}}$ to find the slope.

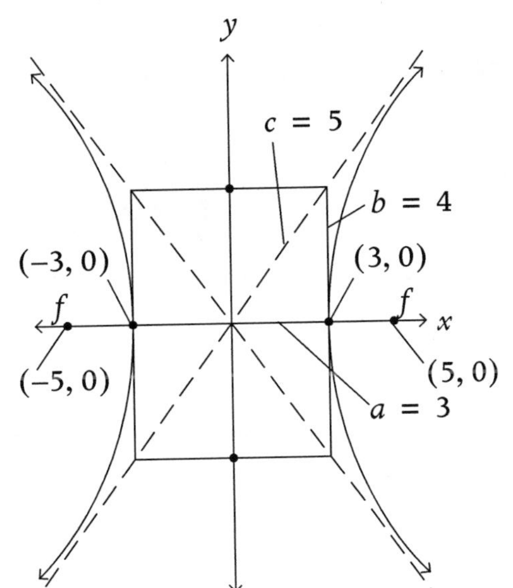

Vertices: $(3, 0), (-3, 0)$.

Asymptotes: $y = \pm\frac{4}{3}x$.

Foci: $(\pm 5, 0)$

$c = \sqrt{a^2 + b^2} = \sqrt{3^2 + 4^2}$

$c = \sqrt{25} = 5$.

π TRY IT

Practice Activities

1. Find the equation of a line, with integral slope, passing through $(2, 2)$, that makes a triangle with an area 9 units^2 with the coordinate axes. Express the answer in the form of $Ax + By = C$.

2. Find the equation of the ellipse having x-intercepts $(6, 0)$ and $(-6, 0)$ and foci $(3, 0)$ and $(-3, 0)$. Write your answer in the form of $\frac{x^2}{a^2} + \frac{y^2}{b^2} = 1$.

3. Given a parabola in the form of $y = ax^2 + bx + c$ that passes through the point $(4, 3)$ and has a vertex $V(6, -3)$, find the coefficient of the x^2 term.

4. Find the distance between the two lines $3x + 4y = 12$ and $3x + 4y = 15$.

5. Find the distance between the centers of the two circles $x^2 + 4x + y^2 - 6y = 2$ and $x^2 + y^2 + 8y = 7$.

6. Given the ellipse $9x^2 + 16y^2 = 144$, find the length of the segment joining a vertex to the nearest focus.

7. Find the equations of the asymptotes of the hyperbola for which the equation is $25y^2 - 16x^2 = 100$. Express your answer in the $y = mx + b$ form.

8. The ellipse $25(x - 2)^2 + 4(y + 3)^2 = 100$ is inscribed in a rectangle, the sides of which are parallel to the coordinate axes. What is the area of the rectangle in square units?

Answer Key

Exponents and Radicals

1. Find all positive values of x that satisfy $x \cdot \sqrt[x]{x^2} = x^x$.

   $^*x = 1$ by observation. Use the rule $\sqrt[r]{b^p} = b^{\frac{p}{r}}$, $x \cdot x^{\frac{2}{x}} = x^x$, and use the rule $x^a \cdot x^b = x^{a+b}$, and set exponents equal to each other.

   If $b^x = b^y$, then $x = y$.

   $\therefore 1 + \frac{2}{x} = x \Rightarrow x + 2 = x^2;\ x^2 - x - 2 = 0$; factor, $(x-2)(x+1) = 0 \Rightarrow x = 2, -1$,

   reject -1, and from $^*x = 2, 1$.

2. If $3 + \sqrt{b}$ is equivalent to the reciprocal of its conjugate, what is the value of b?

   The conjugate of $3 + \sqrt{b}$ is $3 - \sqrt{b}$ and its reciprocal is $\frac{1}{(3 - \sqrt{b})}$.

   $\therefore 3 + \sqrt{b} = \frac{1}{(3 - \sqrt{b})}$, multiplying both sides by the denominator,

   $(3 + \sqrt{b})(3 - \sqrt{b}) = 9 - b$.

   $\therefore 9 - b = 1$.

   $\therefore b = 8$.

3. If a and b are positive and $\sqrt{(a+2)(b+3)} = \sqrt{2b} + \sqrt{3a}$, find the product ab.

   Expand the radicand and square both sides.

   $(\sqrt{ab + 2b + 3a + 6})^2 = (\sqrt{2b} + \sqrt{3a})^2 \Rightarrow ab + 2b + 3a + 6 = 2b + 2\sqrt{6ab} + 3a$.

   Combining similar terms $ab - 2\sqrt{6ab} + 6 = 0$, factor

   $(\sqrt{ab} - \sqrt{6})^2 = 0 \Rightarrow \sqrt{ab} - \sqrt{6} = 0 \Rightarrow \sqrt{ab} = \sqrt{6}$; square both sides.

   The product ab is 6.

   Notice we didn't find a or b. Always keep the question or answer form in mind.

   Also, we know that the factors of $a^2 - 2ab + b^2$ are $(a-b)^2$, so the factors of $ab - 2\sqrt{6ab} + 6$ are $(\sqrt{ab} - \sqrt{6})^2$. This will appear again in a later unit.

4. Find $p + q$ in terms of a, b, and c, where $a > b$.

   $$\left(\frac{x^b y^a x^a}{y^b}\right)^c = x^p y^q.$$

   $$\left(\frac{x^b y^a x^a}{y^b}\right)^c = \left(\frac{x^{b+a} y^a}{y^b}\right)^c = \frac{x^{bc+ac} y^{ac}}{y^{bc}} = x^{bc+ac} y^{ac-bc}.$$

$x^{bc+ac}y^{ac-bc} = x^p y^q \Rightarrow p+q = bc+ac+ac-bc.$

$\therefore p+q = 2ac.$

5. For the expression $\dfrac{(\sqrt[3]{9}\cdot\sqrt[4]{4})}{\sqrt[6]{12}} = \sqrt[m]{a}$, where a is a whole number and m is the smallest possible root, find the sum $a+m$.

Solution 1:

Change each radical to similar roots.

$\sqrt[3]{9} = \sqrt[6]{81}$ because $\sqrt[3]{9} = (3^2)^{\frac{1}{3}} = 3^{\frac{2}{3}}$, $\sqrt[6]{81} = (3^4)^{\frac{1}{6}} = 3^{\frac{2}{3}}$, and $\sqrt[4]{4} = \sqrt{2} = \sqrt[6]{8}$, because $\sqrt[4]{4} = (2^2)^{\frac{1}{4}} = 2^{\frac{1}{2}} = \sqrt{2}$ and $\sqrt[6]{8} = (2^3)^{\frac{1}{6}} = 2^{\frac{1}{2}} = \sqrt{2}$.

$\therefore \dfrac{\sqrt[6]{81}\cdot\sqrt[6]{8}}{\sqrt[6]{12}} = \sqrt[6]{\dfrac{81\cdot 8}{12}} = \sqrt[6]{54}.$

$\therefore \sqrt[m]{a} = \sqrt[6]{54}$, so the sum $a+m = 54+6 = 60$.

Solution 2:

$\sqrt[3]{9} = 3^{\frac{2}{3}}$ from solution 1 above, and $\sqrt[4]{4} = 2^{\frac{1}{2}}$.

$\therefore 3^{\frac{2}{3}}\cdot 2^{\frac{1}{2}} = 3^{\frac{4}{6}}\cdot 2^{\frac{3}{6}} = (3^4\cdot 2^3)^{\frac{1}{6}} = (81\cdot 8)^{\frac{1}{6}}$ and $\sqrt[6]{12} = 12^{\frac{1}{6}}$.

$\left(\dfrac{81\cdot 8}{12}\right)^{\frac{1}{6}}$, etc.

6. Simplify: $\dfrac{\left(9^x\cdot 3^2\cdot \dfrac{1}{3^{-x}}\right) - 27^x}{3^{3x}\cdot 9}$.

$\dfrac{(3^{2x}\cdot 3^2\cdot 3^x) - 3^{3x}}{3^{3x}\cdot 3^2} = \dfrac{3^{3x}\cdot 3^2 - 3^{3x}}{3^{3x}\cdot 3^2}$; factor out a 3^{3x}, then cancel them.

$\dfrac{3^{3x}(3^2-1)}{3^{3x}\cdot 3^2} = \dfrac{8}{9}$. $\therefore$ The answer is $\dfrac{8}{9}$.

7. The square of $a+b\sqrt{3}$ is $84-30\sqrt{3}$. Find all ordered pairs of (a, b), where a and b are integers.

$(a+b\sqrt{3})^2 = 84-30\sqrt{3} \Rightarrow a^2+2ab\sqrt{3}+3b^2 = 84-30\sqrt{3}$

Set corresponding terms equal, $a^2+3b^2 = 84$ and $2ab\sqrt{3} = -30\sqrt{3} \Rightarrow {}^*ab = -15$, or $a = \dfrac{-15}{b}$; substitute in the previous equation, $\left(-\dfrac{15}{b}\right)^2 + 3b^2 - 84 = 0$, multiply by b^2, $225 + 3b^4 - 84b^2 = 0$; divide by $3 \Rightarrow b^4 - 28b^2 + 75 = 0$.

Factor, $(b^2 - 3)(b^2 - 25) = 0 \Rightarrow b = \pm 5, \pm\sqrt{3}$; reject $\sqrt{3}$ because it is not an integer.

$\therefore b = \pm 5$; if $b = 5$, then $a = -3$, and if $b = -5$, then $a = 3$.

Answers are $(3, -5)$, $(-3, 5)$.

Notice at *, because $ab = -15$ and a and b are integers, a must equal ± 3 or ± 5. You should substitute to find the solutions.

8. Simplify: $\dfrac{\sqrt{2}}{\sqrt{2} + \sqrt{3} - \sqrt{5}}$.

Rationalize as shown: $\dfrac{\sqrt{2}}{(\sqrt{2} + \sqrt{3}) - \sqrt{5}} \cdot \dfrac{\sqrt{2} + \sqrt{3} + \sqrt{5}}{(\sqrt{2} + \sqrt{3}) + \sqrt{5}}$

$\dfrac{2 + \sqrt{6} + \sqrt{10}}{2 + 3 + 2\sqrt{6} - 5} \cdot \dfrac{2 + \sqrt{6} + \sqrt{10}}{2\sqrt{6}} \cdot \dfrac{\sqrt{6}}{\sqrt{6}}$; rationalize again.

$\dfrac{2\sqrt{6} + 6 + 2\sqrt{15}}{12} = \dfrac{\sqrt{6} + \sqrt{15} + 3}{6}$ (Answer).

9. Find all real value(s) of x for which the following is true:

$\dfrac{x + \sqrt{x + 1}}{x - \sqrt{x + 1}} = \dfrac{5}{11}$.

The following is a common strategy because it reduces the amount of work that you do.

Let $a^2 = x + 1$, then $x = a^2 - 1$. Substitute into $\dfrac{(x + \sqrt{x + 1})}{(x - \sqrt{x + 1})} = \dfrac{5}{11}$ and get

$\dfrac{(a^2 - 1 + a)}{(a^2 - 1 - a)} = \dfrac{5}{11}$, cross multiply, $11a^2 + 11a - 11 = 5a^2 - 5a - 5$; combine and factor

$6a^2 + 16a - 6 = 0$, $2(3a^2 + 8a - 3) = 0$; divide by 2, $(3a - 1)(a + 3) = 0 \Rightarrow a = \dfrac{1}{3}$,

-3. Substitute in for $a^2 = x + 1$, $\dfrac{1}{9} = x + 1 \Rightarrow x = -\dfrac{8}{9}$, and

$(-3)^2 = x + 1 \Rightarrow 9 = x + 1$, $x = 8$.

However, $x = 8$ doesn't check, so $-\dfrac{8}{9}$ is the only answer.

Polynomials and Factoring

1. Write as a product of prime factors: $(a + b)(a - b) - c(c + 2b)$.

Expand: $a^2 - b^2 - c^2 - 2bc$; group the last three terms,

$a^2 - (b^2 + 2bc + c^2) = a^2 - (b + c)^2 = (a + b + c)(a - b - c)$.

2. Factor completely: $4a^2 - 9b^2 + 20a + 24 - 6b$.

   If you rewrite 24 as $25 - 1$, there are two perfect trinomials; group them.

   $4a^2 + 20a + 25 - 9b^2 - 6b - 1 = (4a^2 + 20a + 25) - (9b^2 + 6b + 1)$.

   Factor: $(2a + 5)^2 - (3b + 1)^2$

   $(2a + 5 + 3b + 1)(2a + 5 - 3b - 1)$.

3. Find the sum of the factors: $4a^2 + 4ab + b^2 - 2a - b - 12$.

   The first three terms are a perfect trinomial square, the next two terms are the middle term, and the -12 is the constant term.

   $(2a + b)^2 - (2a + b) - 12$; let $x = 2a + b$, then $x^2 - x - 12$.

   Factor and then substitute $2a + b$ for x.

   $(x - 4)(x - 3) = (2a + b - 4)(2a + b + 3)$.

   $\therefore$ The sum of the factors is $4a + 2b - 1$.

4. Find the least common multiple (LCM) of A, B, and C, where

   $A = 12x^5 + 24x^4 + 12x^3$; $B = 15x^4 - 15x^2$; $C = 9x^6 - 18x^5 + 9x^4$.

   Factor A, B, and C:

   A: $12x^5 + 24x^4 + 12x^3 = 12x^3(x^2 + 2x + 1) = 12x^3(x + 1)^2$.

   B: $15x^4 - 15x^2 = 15x^2(x^2 - 1) = 15x^2(x + 1)(x - 1)$.

   C: $9x^6 - 18x^5 + 9x^4 = 9x^4(x^2 - 2x + 1) = 9x^4(x - 1)^2$.

   The LCM of $12x^3$, $15x^2$, and $9x^4$ is $180x^4$.

   The LCM of $(x + 1)^2$, $(x + 1)(x - 1)$, and $(x - 1)^2$ is $(x + 1)^2(x - 1)^2$.

   The answer is $180x^4(x - 1)^2(x + 1)^2$.

5. Factor completely: $yx^2 - x^2z + y^2z - yz^2 + xy^2 - xz^2$.

   Group terms: $(yx^2 - x^2z) + (y^2z - yz^2) + (xy^2 - xz^2)$; factor out the common factors, $x^2(y - z) + yz(y - z) + x(y + z)(y - z)$; factor out the common factor of $y - z$, $(y - z)[x^2 + yz + x(y + z)] = (y - z)(x^2 + xz + xy + yz)$; group inside, $(y - z)[x(x + z) + y(x + z)]$; factor out $(x + z)$.

   $\therefore$ The answer is $(y - z)(x + z)(x + y)$.

6. Factor completely: $x^2 - 2xy + y^2 + 10x - 10y + 9$.

   Group the first three terms and $10x - 10y$, $(x^2 - 2xy + y^2) + (10x - 10y) + 9$; factor, $(x - y)^2 + 10(x - y) + 9$. Let $a = (x - y)$; this makes the problem easier. Substitute and factor so $a^2 + 10a + 9 = (a + 9)(a + 1)$ and substitute back $(x - y)$ for a and get $(x - y + 9)(x - y + 1)$, which is the answer.

7. Factor completely: $16a^{x+2} - 20a^{x+3} + 4a^{x+4}$.

   Solution 1:

   Rewrite the expression as $16a^x \cdot a^2 - 20a^x \cdot a^3 + 4a^x \cdot a^4$.

   Factor out $4a^x a^2$. $\therefore$ $4a^x a^2(4 - 5a + a^2) = 4a^x a^2(a - 4)(a - 1)$.

   The answer is $4a^{x+2}(a - 4)(a - 1)$.

   Solution 2:

   Factor out $4a^{x+2}$ immediately.

8. Factor completely:

   $x^4 + (3 + t)x^2 - 2t - 10$.

   Expand, $x^4 + 3x^2 + tx^2 - 2t - 10$.

   Group $(x^4 + 3x^2 - 10) + (tx^2 - 2t)$; factor $(x^2 + 5)(x^2 - 2) + t(x^2 - 2)$; factor out $(x^2 - 2)$.

   The answer is $(x^2 - 2)(x^2 + 5 + t)$.

Rational Expressions

1. For what integer value(s) of x does $\frac{(35 - 17x + 2x^2)}{(25 - x^2)}$ represent a perfect square of some rational number?

   $\frac{35 - 17x + 2x^2}{25 - x^2}$; factor and cancel $\frac{(2x - 7)(x - 5)}{(5 + x)(5 - x)}$; $\frac{(x - 5)}{(5 - x)} = -1$. $\therefore$ 5 is not a solution, as 5 makes the denominator and numerator equal to zero, so we have $-\frac{2x - 7}{x + 5}$ or $\frac{7 - 2x}{x + 5}$.

   The expression $\frac{(7 - 2x)}{(x + 5)}$ must equal a perfect square (number must be positive); the expression is positive for $x = -4, -3, -2, -1, 0, 1, 2, 3$.

   Substituting these, the only one that works is -1, which equals $\frac{9}{4}$.

   $\therefore$ The answer is -1.

2. If $\frac{(n - 1) + n + (n + 1)}{n} = \frac{2{,}002 + x + 2{,}004}{x}$ and $n \neq 0$, find x.

   Simplify both numerators, $\frac{3n}{n} = \frac{4{,}006 + x}{x}$; n's cancel and cross-multiply,

   $3x = 4{,}006 + x \Rightarrow x = 2{,}003$.

3. If $\frac{x^2+3xy-4y^2}{x^2-y^2} = \frac{13}{7}$, $x \neq y$, find the ratio of y to x.

   Factor, $\frac{(x+4y)(x-y)}{(x+y)(x-y)}$; cancel the $x-y$'s and $\frac{(x+4y)}{(x+y)} = \frac{13}{7}$; cross-multiply,

   $7x + 28y = 13x + 13y$; simplify to $15y = 6x$.

   Answer form is $\frac{y}{x}$, so divide by $15x$ to isolate the y and create an x in the denominator,

   $\frac{15y}{15x} = \frac{6x}{15x}$; reduce and cancel. $\therefore \frac{y}{x} = \frac{2}{5}$.

4. Let $f(x) = x^2$ and $g(x) = 3x + 10$, simplify the following expression:

   $\frac{f(x+3)-g(x+3)}{(x+3)+2}$.

   $\frac{f(x+3)-g(x+3)}{(x+3)+2} = \frac{(x+3)^2-[3(x+3)+10]}{x+5} = \frac{x^2+6x+9-3x-9-10}{x+5}$;

   simplify numerator and then factor it,

   $\frac{x^2+3x-10}{x+5} = \frac{(x+5)(x-2)}{x+5}$.

   $\therefore$ Cancel the $x+5$'s and the answer is $x-2$.

5. Solve for x: $\dfrac{\frac{x}{x+1}-\frac{1-x}{x}}{\frac{x}{1+x}+\frac{1-x}{x}} = 1$.

   Simplify numerator and denominator by multiplying as shown:

   $$\frac{\frac{x}{x}\cdot\frac{x}{x+1}-\frac{1-x}{x}\cdot\frac{x+1}{x+1}}{\frac{x}{x}\cdot\frac{x}{1+x}+\frac{1-x}{x}\cdot\frac{x+1}{x+1}} = 1 \Rightarrow \frac{\frac{x^2-1+x^2}{x(x+1)}}{\frac{x^2+1-x^2}{x(x+1)}} = \frac{\frac{2x^2-1}{x(x+1)}}{\frac{1}{x(x+1)}} = 1.$$

   Now multiply numerator and denominator by $x(x+1)$ and we have

   $2x^2 - 1 = 1 \Rightarrow 2x^2 = 2$, so $x = \pm 1$; reject -1 because the denominator will equal 0.

   $\therefore x = 1$.

6. If $A + B = 10$ and $A - B = 6$, what is the value of $\frac{3A^2+6AB+3B^2}{5A^2-5B^2}$?

   Factor $\frac{3A^2+6AB+3B^2}{5A^2-5B^2} = \frac{3(A^2+2AB+B^2)}{5(A^2-B^2)} = \frac{3(A+B)^2}{5(A+B)(A-B)}$,

   cancel $(A + B)$'s to get

$\frac{3(A+B)}{5(A-B)}$; substitute values given in the problem so $\frac{3 \cdot 10}{5 \cdot 6} = 1$.

You could solve this problem by solving the system of equations to get $A = 8$ and $B = 2$, substituting these values in the problem, and using a calculator. You should know both methods.

7. For how many integral value(s) of x does $\frac{(3x+17)}{(x+2)}$ have an integral value?

   This is similar to solving a Diophantine Equation (an equation for which the solutions are integers).

   Divide $\frac{(3x+17)}{(x+2)}$ and get $3 + \frac{11}{(x+2)}$; x must be an integer for $\frac{11}{(x+2)}$ to be an integer and $x + 2$ divides 11 only if $x + 2 = \pm 11, \pm 1$.

   $x = 9, -13, -3, -1$.

   $\therefore$ There are four values.

8. Find the value(s) of constants C and D and write the answer in the form (C, D):

   $\frac{C}{x-2} + \frac{D}{x+1} = \frac{6x}{x^2 - x - 2}$.

   Multiply $\frac{C}{x-2} + \frac{D}{x+1} = \frac{6x}{x^2 - x - 2}$ by $(x-2)(x+1)$ and get

   $C(x+1) + D(x-2) = 6x \Rightarrow Cx + C + Dx - 2D = 6x$; create a system of equations by setting coefficients of x equal to 6 and C and D equal to 0.

   $\therefore C + D = 6$.

   $C - 2D = 0$; subtracting, $3D = 6 \Rightarrow D = 2$ and $C = 4$.

   $\therefore (C, D) = (4, 2)$.

9. The rational expression $\frac{(x^2 + ax + b)}{(x^2 - 2ax + 2b)}$, with a and b constants, reduces to $\frac{(x+3)}{(x+6)}$ with certain restrictions on x. Find the value of b.

   Because $(x + 3)$ is a factor of the numerator and $(x + 6)$ is a factor of the denominator, first let $x = -3$ and substitute -3 in the numerator.

   $\therefore 9 - 3a + b = 0$; then let $x = -6$, and substitute it in the denominator and get $36 + 12a + 2b = 0$. Then solve the system of equations.

   1) $9 - 3a + b = 0$

   2) $36 + 12a + 2b = 0$; multiply equation 1 by 4, and add equations together

   $$\begin{array}{r} 36 - 12a + 4b = 0 \\ \underline{36 + 12a + 2b = 0} \\ 72 + 6b = 0 \Rightarrow b = -12. \end{array}$$

Again, we only need to find the value of b, which is -12.

Complex Numbers

1. Evaluate $(1-i)^{-8}$.

   $(1-i)^{-8} = [(1-i)^2]^{-4}$ and $(1-i)^2 \Rightarrow (-2i)$.

   $\therefore (-2i)^{-4} = \frac{1}{16}i^{-4} = \frac{1}{16}$.

2. Solve for x in the domain of complex numbers.

   $5x + 2 = ix - 4i$

   $5x - ix = -2 - 4i \Rightarrow x(5-i) = -2-4i$

   $x = -\frac{(2+4i)}{(5-i)}$, rationalize by using the conjugate of $5-i$,

   $x = \frac{-(2+4i)}{5-i} \cdot \frac{5+i}{5+i} = -\frac{10+22i+4i^2}{25-i^2}$

   $x = \frac{-6-22i}{26} = -\frac{3}{13} - \frac{11}{13}i$.

3. If $Z = 3+2i$, determine the value of $Z^4 - Z^3 + Z^2 - Z + 1$.

   Express your answer in the form $a + bi$.

   $Z^4 - Z^3 + Z^2 - Z + 1$, substitute $3+2i$ for Z,

   $(3+2i)^2[(3+2i)^2 - (3+2i) + 1] - (3+2i) + 1$

   $(3+2i)^2 = 9 + 12i + 4i^2 = 5 + 12i$

   $\therefore (5+12i)(5+12i-3-2i+1) - 3 - 2i + 1$

   $= (5+12i)(3+10i) = 15 + 86i + 120i^2$

   $= -105 + 86i - 2i - 2 = -107 + 84i$, which is the answer.

4. Find the square root of $15 - 8i$.

   Let $a + bi = \sqrt{15-8i}$ and square both sides.

   $(a+bi)^2 = (\sqrt{15-8i})^2 \Rightarrow a^2 + 2abi + b^2i^2 = 15 - 8i$

   $b^2i^2 = -b^2$, so $a^2 + 2abi - b^2 = 15 - 8i$; create a system of equations by setting the real number parts equal, $a^2 - b^2 = 15$, and the complex number parts equal,

   $2abi = -8i \Rightarrow ab = -4$ and $a = -\frac{4}{b}$; substitute for a and obtain

   $\left(-\frac{4}{b}\right)^2 - b^2 = 15 \Rightarrow \frac{16}{b^2} - b^2 = 15$; multiplying by b^2, we get

   $16 - b^4 = 15b^2 \Rightarrow b^4 + 15b^2 - 16 = 0$; then factor

$(b^2 + 16)(b^2 - 1) = 0 \Rightarrow b^2 = -16,\ b^2 = 1$; reject roots of $b^2 = -16$ because b is a real number, so $b^2 = 1 \Rightarrow b = \pm 1$.

Because $a = -\frac{4}{b}$, if $b = 1$, then $a = -4$; and if $b = -1$, then $a = 4$.

The answers written in the form of $a + bi$ are $-4 + i$ and $4 - i$.

5. Solve: $ix^2 + 9x = 20i$.

   Rewrite and use the quadratic formula: $ix^2 + 9x - 20i = 0$.

   $$\frac{-9 \pm \sqrt{81 - 4(i)(-20i)}}{2i} = \frac{-9 \pm \sqrt{81 - 80}}{2i} = \frac{-9 \pm 1}{2i}$$

   $$= \frac{-9 + 1}{2i} = \frac{-8}{2i} \cdot \frac{i}{i} = \frac{-8i}{2i^2},\ 2i^2 = -2,\ \therefore \frac{-8i}{-2} = 4i$$

   and $\frac{-9 - 1}{2i} = \frac{-10}{2i} \cdot \frac{i}{i} = 5i$.

   $\therefore x = 4i, 5i$.

6. Find the complex number Z for which the following equation is true: $Z + 6z = 7 + 3i$. Z is a complex number and z is its conjugate.

   Let $Z = a + bi$, then $z = a - bi$. Substitute these in the original equation.

   $\therefore a + bi + 6(a - bi) = 7 + 3i \Rightarrow a + bi + 6a - 6bi = 7 + 3i$; combine the like terms on the side, $7a - 5bi = 7 + 3i$; set the real-number parts equal, $7a = 7 \Rightarrow a = 1$; then set the imaginary parts equal, $-5b = 3 \Rightarrow b = -\frac{3}{5}$.

   $\therefore Z = 1 - \frac{3}{5}i$.

7. Find all ordered pairs (x, y) of real numbers that satisfy the equation $(x^2 - x - 5) + i(y - 12) = 1 - 7i$.

   Use the definition of equal complex numbers, $a + bi = c + di$ if $a = c$ and $b = d$.

   $\therefore x^2 - x - 5 = 1 \Rightarrow x^2 - x - 6 = 0$ and $y - 12 = -7$.

   Factor, $(x - 3)(x + 2) = 0 \Rightarrow x = 3, -2$ and $y = 5$.

   $\therefore$ The answers are $(3, 5), (-2, 5)$.

8. Evaluate: $(i^{(3^3)}) - (i^3)^3$.

   The answer is not zero. $a^{(b^c)}$ means that you first raise b to the c power, then raise a to that power. Example, $2^{(2^3)} = 2^8$.

   $\therefore i^{(3^3)} - (i^3)^3 = i^{27} - i^9 = (i^4)^6 \cdot i^3 - (i^4)^2 \cdot i$.

$(i^4) = 1$, so the problem reduces to $i^3 - i \Rightarrow -i - i = -2i$.

$\therefore\ i^{(3^3)} - (i^3)^3 = -2i$.

Polynomial Functions

1. Find the polynomial function $P(x)$ for which the zeros are ± 1, 3 and $P(0) = 9$.
   Because there are three roots, set up a cubic equation where k represents the leading coefficient. Substitute the zeros for x to find k.
   $P(x) = k(x+1)(x-1)(x-3)$ and because $P(0) = 9$,
   $9 = k(0+1)(0-1)(0-3) \Rightarrow 3k = 9, k = 3$.
   $\therefore\ P(x) = 3(x+1)(x-1)(x-3)$; expand, $3(x^2-1)(x-3)$.
   $P(x) = 3x^3 - 9x^2 - 3x + 9$.

2. If 2 is a root of $4x^3 - 3x^2 - kx - 4k^2 = 0$, find the values of k.
   Because 2 is a root by the Factor Theorem, then
   $4(2)^3 - 3(2)^2 - 2k - 4k^2 = 32 - 12 - 2k - 4k^2 = 0$; factor, $2k^2 + k - 10 = 0$,
   $(2k+5)(k-2) = 0$.
   $\therefore\ k = -\frac{5}{2}, 2$.

3. $P(x) = 3x^3 - 6x^2 - 3x + 6$; for which values of x is $P(x) > 0$?
   $P(x) = 3x^3 - 6x^2 - 3x + 6$; factor by grouping,
   $P(x) = 3x^2(x-2) - 3(x-2) \Rightarrow (3x^2-3)(x-2)$;
   factor $P(x) = 3(x+1)(x-1)(x-2)$.
   Show ± 1, 2 as the zero points on a number line. Find the regions where $P(x)$ is positive.
   $\therefore\ P(x) > 0;\ -1 < x < 1$ or $x > 2$.

4. Find $P(x)$ such that $P(1) = 6$ and $P(x) > 0$ only when $x < 0$ and $2 < x < 3$.
   As the zeros of the function are 0, 2, 3, form the factors,
   $P(x) = k(x)(x-2)(x-3)$.
   Because $P(1) = 6$, substitute and simplify, $6 = k(1)(-1)(-2) \Rightarrow 2k = 6, k = 3$.
   $\therefore\ P(x) = 3x(x-2)(x-3)$; expand.
   The answer is $P(x) = 3x^3 - 15x^2 + 18x$.

5. If $f(x) = x^4 + x^3 + ax + b$, find a and b such that $f(2) = -20$ and $f(-3) = 15$.
   Substitute for x and $f(x)$ to form a system of equations,
   $-20 = 16 + 8 + 2a + b \Rightarrow 2a + b = -44$
   $15 = 81 - 27 - 3a + b \Rightarrow -3a + b = -39$.

Subtract the two equations, $5a = -5$.

$\therefore\ a = -1$ and $b = -42$.

6. Given the polynomial function $f(x) = 2x^2 + ix - 1,\ i = \sqrt{-1}$; what is the remainder if $f(x)$ is divided by $x - i$?

   Solution 1:

   Using The Remainder Theorem, which states when a polynomial $f(x)$ is divided by $x - a$, the remainder is $f(a)$.

   $\therefore\ f(i) = 2(i)^2 + i(i) - 1 = -2 - 1 - 1 = -4.$

   Solution 2:

   Use Synthetic Division:

   $$\begin{array}{r|rrr} i & 2 & i & -1 \\ & & 2i & -3 \\ \hline & 2 & 3i & -4 \end{array}$$

   $\therefore\ f(i) = -4.$

7. Given the graph of a cubic function, $y = f(x)$ with zeros at 3 and 0, and $f(2) = 4$, find $f(4)$.

   As seen on the graph, 0 is a double root, so the function is of the form $f(x) = kx^2(x - 3)$.

   Substitute the ordered pair $(2, 4)$, $4 = 4k(2 - 3)$; solve and $k = -1$.

   $\therefore\ f(x) = -1(x^2)(x - 3)$ and

   $f(4) = -16(4 - 3) = -16.$

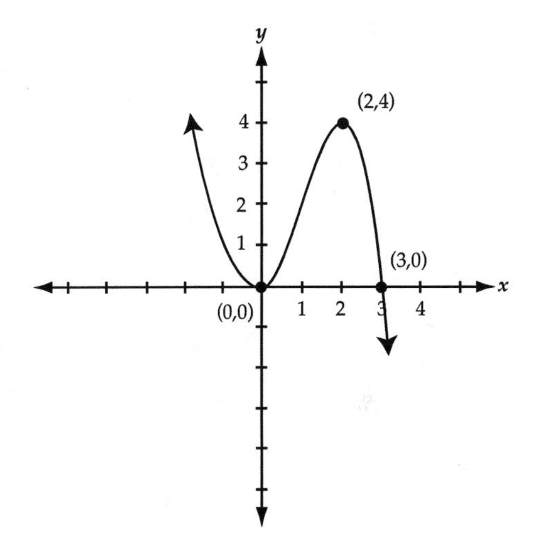

8. If $3x + 1$ is a factor of $6x^2 - kx - 1$, find k.

   Solution 1:

   Because $(3x + 1)$ is one of the factors, the other one must be $(2x - 1)$.

   $\therefore\ (3x + 1)(2x - 1) = 6x^2 - kx - 1$ as $2x \cdot 3x = 6x^2$ and $1(-1) = -1$.

   $(3x - 1)(2x - 1) = 6x^2 - x - 1$, so $k = 1$.

   Solution 2:

   Using synthetic division requires more work but you should know both methods.

   Because $(3x + 1)$ is a factor, $-\dfrac{1}{3}$ is a root; use the Remainder Theorem.

$$\begin{array}{r|ccc} -\frac{1}{3} & 6 & -k & -1 \\ & & -2 & \frac{1}{3}k + \frac{2}{3} \\ \hline & 6 & -2-k & \frac{1}{3}k - \frac{1}{3} \end{array}$$

$\therefore$ Solve $\frac{1}{3}k - \frac{1}{3} = 0 \Rightarrow k = 1.$

Inverse Functions, Compositions, and Functions in General

1. If $f(x) = 3x + 2$ and $g(x) = x^2 - 1$, find $g[f^{-1}(-2)]$.

   Find the inverse of $y = 3x + 2$; interchange x and y,

   $x = 3y + 2$; and solve for y, $f^{-1}(x) = \frac{(x-2)}{3}$. $\therefore f^{-1}(-2) = -\frac{4}{3}$.

   Then $g\left[-\frac{4}{3}\right] = \frac{16}{9} - 1 = \frac{7}{9}$.

   $\therefore g[f^{-1}(-2)] = \frac{7}{9}$.

2. If $f(x) = \frac{[x(x-3)]}{2}$, find the positive integer n such that $f(n) = 189$.

   $\therefore f(n) = \frac{n(n-3)}{2} \Rightarrow \frac{n(n-3)}{2} = 189$; multiply by 2,

   $n(n-3) = 2 \cdot 189 = 378 \Rightarrow n^2 - 3n - 378 = 0$; factor, $(n+18)(n-21) = 0$;
   $n = 21$ because you want the positive integer.

3. If $f(x) = x + 1$, $g(x) = 2x + 1$, and $h(x) = 3x + 1$, for what x does
   $3f(g(x)) + 2g(h(x)) = h(g(x))$?

   $3f(2x+1) + 2g(3x+1) = h(2x+1)$

   $3[(2x+1)+1] + 2[2(3x+1)+1] = 3(2x+1)+1$

   $6x + 6 + 12x + 6 = 6x + 4 \Rightarrow 12x = -8$ ($6x$'s cancel).

   $\therefore x = -\frac{2}{3}$.

4. If $f(x-7) = 3x + 2a$, where a is a real number and $f(a) = 17$, find the numerical value of a.

   $f(x-7) = 3x + 2a$ and $f(a) = 17$; then let $t = x - 7 \Rightarrow x = t + 7$.

$\therefore f(t) = 3(t+7) + 2a = 3t + 21 + 2a$

and $f(a) = 3a + 21 + 2a = 5a + 21 \Rightarrow 17 = 5a + 21.$

Solve, and the answer is $a = -\frac{4}{5}$.

5. If $f(n)$ is a function such that $f(1) = f(2) = f(3) = 2$ and

   $f(n+2) = \dfrac{f(n+1) \cdot f(n) + 1}{f(n-1)}$ for $n \geq 3$, find the numerical value of $f(6)$.

   $f(6) = f(4+2) = \dfrac{f(5) \cdot f(4) + 1}{f(3)}$

   $f(5) = f(3+2) = \dfrac{f(4) \cdot f(3) + 1}{f(2)}$

   $f(4) = f(2+2) = \dfrac{f(3) \cdot f(2) + 1}{f(1)}$

   $f(4) = \dfrac{2 \cdot 2 + 1}{2} = \dfrac{5}{2}$ and $f(5) = \dfrac{\frac{5}{2} \cdot 2 + 1}{2} = 3.$

   $\therefore f(6) = \dfrac{3 \cdot \frac{5}{2} + 1}{2} = \dfrac{\frac{17}{2}}{2} = \dfrac{17}{4}.$

6. A function f has the following three properties for all positive integers for n:
   $f(1) = 1$, $f(2) = 3$, and $f(n) = 2f(n-1) - f(n-2)$ for $n \geq 3$. What is the value of $f(100)$? Given $f(n) = 2f(n-1) - f(n-2) \Rightarrow f(3) = 2f(2) - f(1)$.

   $\therefore f(3) = 2 \cdot 3 - 1 = 5$. $f(4) = 2f(3) - f(2) \Rightarrow 2 \cdot 5 - 3 = 7$.

   By the above pattern, $f(3) = 5$, $f(4) = 7$, etc.

   Then $f(n) = 2n - 1$.

   $\therefore f(100) = 2 \cdot 100 - 1 = 199.$

7. If $p(x) = x^2 + 8$ and $q(x) = 2x$, for what value(s) of x is $p(q(x)) = q(p(x))$?

   $p(2x) = (2x)^2 + 8 = 4x^2 + 8$

   $q(x^2 + 8) = 2(x^2 + 8) = 2x^2 + 16$

   $\therefore p(q(x)) = q(p(x)) \Rightarrow 4x^2 + 8 = 2x^2 + 16$

   $2x^2 = 8 \Rightarrow x^2 = 4.\ \therefore x = \pm 2.$

8. Given two functions f and g, $g(x) = 3x + 2$ and f is unknown. If $f(g(x)) = x^2 - x - 3$, what is the value of $f(1)$?

   For what value of x is $3x + 2 = 1$? Solve and $x = -\frac{1}{3}$.

Substituting $-\frac{1}{3}$ for x in $x^2 - x - 3$, we get

$$\left(-\frac{1}{3}\right)^2 - \left(-\frac{1}{3}\right) - 3 = \frac{1}{9} + \frac{3}{9} - 3 = \frac{4}{9} - \frac{27}{9} = -\frac{23}{9}.$$

$\therefore f(1) = -\frac{23}{9}$.

9. If $g(x) = -\frac{2}{3}(x+5)$, then find $g^{-1}(-10)$.

   Let $g^{-1}(-10) = w$ (an arbitrary value), then $g(w) = -10$.

   $\therefore -10 = -\frac{2}{3}(w+5) \Rightarrow -30 = -2(w+5)$, $-30 = -2w - 10$, $w = 10$.

   $\therefore g^{-1}(-10) = 10$.

Finding Zeros of Functions

1. The product of two of the roots of the equation $x^3 + 2x + 12 = 0$ is 6. Find the roots.
   Let r_1, r_2, and r_3 represent the roots. If the degree of a polynomial is odd,

   then$r_1 \cdot r_2 \cdot r_3 = -\frac{c}{a}$. Using this rule, the product of the roots is -12, and because it was

   given that the product of two of the roots is 6, then one of the roots is -2.
   Using synthetic division:

-2	1	0	2	12
		-2	4	-12
	1	-2	6	0

   The reduced or depressed equation is $x^2 - 2x + 6 = 0$. Complete the square to find the other two roots.

   $x^2 - 2x = -6$; take $\frac{1}{2}$ of the middle term, so $\frac{1}{2}(-2) = -1$; then square $(-1) = 1$;

   add 1 to both sides. $\therefore x^2 - 2x + 1 = -6 + 1$; factor and combine,

   $(x-1)^2 = -5 \Rightarrow x - 1 = \pm i\sqrt{5}$ and $x = 1 \pm i\sqrt{5}$.

   The roots are -2, $1 \pm i\sqrt{5}$.

2. The sum of the squares of the roots of the equation $x^2 + 4hx = 5$ is 154. Compute $|h|$.
   $r_1 + r_2 = -4h$ and $r_1 \cdot r_2 = -5$ (sum and product of roots),
   $(r_1 + r_2)^2 = r_1^2 + 2r_1r_2 + r_2^2$, then we can rewrite $r_1^2 + r_2^2$ as $(r_1 + r_2)^2 - 2r_1r_2$.
   Using $(r_1 + r_2)^2 - 2r_1 \cdot r_2$ and substituting,

$(-4h)^2 - 2(-5) = 16h^2 + 10 = 154$

$16h^2 = 144 \Rightarrow h^2 = 9, h = \pm 3$

and the answer is $|h| = 3$.

3. If the roots of $6x^2 - 13x + 6c = 0$ are r and $\frac{1}{r}$, find the smaller such r.

   Divide by 6, $x^2 - \frac{13}{6}x + c = 0$.

   Because the product of the roots is c, then $c = r \cdot \frac{1}{r} = 1$.

   $\therefore$ The equation is $6x^2 - 13x + 6 = 0$; factor, $(3x - 2)(2x - 3) = 0$;

   and $r = \frac{2}{3}, \frac{3}{2}$ and the smaller root is $\frac{2}{3}$.

4. One root of $4x^3 - 8x^2 + cx + d = 0$ is -1. The other two roots are equal. Find d.

   Because -1 is a root, use synthetic division and the remainder must equal 0.

$$\begin{array}{r|rrrr} -1 & 4 & -8 & c & d \\ & & -4 & 12 & -c-12 \\ \hline & 4 & -12 & c+12 & d-c-12 \end{array}$$

   $\therefore$ Because $d - c - 12 = 0$, $d = c + 12$ (we'll use this later).

   The reduced or depressed equation is $4x^2 - 12x + c + 12 = 0$;

   because there is a double root, the discriminant $b^2 - 4ac = 0$.

   $\therefore$ $(-12)^2 - 4(4)(c + 12) = 0 \Rightarrow 144 - 16c - 192 = 0$,

   $-16c = 48 \Rightarrow c = -3$; using $d = c + 12$,

   $d = -3 + 12, d = 9$.

5. Let $f(x) = \frac{(x+4)^2}{(x-3)}$ and $g(x) = \frac{(2x^2 - 12x - 31)}{(x-3)}$; if $h(x) = f(x) + g(x)$, find the zeros of $h(x)$.

   $f(x) + g(x) = \frac{(x^2 + 8x + 16)}{(x-3)} + \frac{(2x^2 - 12x - 31)}{(x-3)}$.

   Because the rational expressions have the same denominator, add them,

   $h(x) = \frac{x^2 + 8x + 16 + 2x^2 - 12x - 31}{x - 3} = \frac{3x^2 - 4x - 15}{x - 3}$;

   factor the numerator $\frac{(3x+5)(x-3)}{x-3}$;

the zeros are $x = 3, -\frac{5}{3}$, but you can't have $x = 3$ because the denominator would equal 0.

So $x = -\frac{5}{3}$.

6. For what value(s) of k does the equation $2x^2 - 2kx + 5k = 0$ have one root?

   For the quadratic equation to have one root, the value of the discriminant $(b^2 - 4ac)$ must be equal to 0.

   $\therefore b^2 - 4ac = (-2k)^2 - 4(2)(5k) = 0 \Rightarrow 4k^2 - 40k = 0$; factor, $4k(k - 10) = 0$.

   $\therefore k = 0, 10$.

7. Find all positive integral values of B for which the following quadratic equation has two distinct integral roots: $x = \frac{B}{(10 - x)}$.

   Simplifying $x = \frac{B}{(10 - x)}$ to $x^2 - 10x + B = 0$. Let r and s represent the two roots.

   $\therefore$ The sum of the roots, $-\frac{b}{a} = 10$, so $r + s = 10$ and the product of the roots, $\frac{c}{a} = B$, so $rs = B$. Set up a table for r, s, and $B = rs$, where we know that $r + s = 10$.

r	s	$B = rs$
1	9	9
2	8	16
3	7	21
4	6	24
5	5	not distinct

   $\therefore B$ has integral values of $\{9, 16, 21, 24\}$.

8. $4 - \sqrt{3}$ is one root of $2x^2 + bx + c = 0$, where b and c are rational numbers. Find the value of $b + c$.

   Let r and s represent the two roots. If $4 - \sqrt{3}$ is a root, then $4 + \sqrt{3}$ is also a root. Using the sum and product rules for a quadratic equation, we get

   $r + s = (4 + \sqrt{3}) + (4 - \sqrt{3}) = 8$, so $-\frac{b}{a} = 8$.

   $r \cdot s = (4 + \sqrt{3})(4 - \sqrt{3}) = 16 - 3 = 13$, so $\frac{c}{a} = 13$.

$\therefore$ Because $2x^2 + bx + c = 0$ and $-\frac{b}{a} = -\frac{b}{2} = 8$, $b = -16$.

Also, $\frac{c}{a} = \frac{c}{2} = 13$, so $c = 26$.

$\therefore$ The answer for $b + c = -16 + 26 = 10$.

Exponential Functions

1. Solve for m: $3^{m+2} = 3^m + 1{,}944$.

   $\therefore 3^m \cdot 3^2 - 3^m = 1{,}944$, factor out a 3^m,

   $3^m(3^2 - 1) = 1{,}944 \Rightarrow 3^m(8) = 1{,}944$

   $3^m = \frac{1{,}944}{8} = 243 = 3^5 \Rightarrow 3^m = 3^5$,

   $\therefore m = 5$.

2. Find x: $2^{x+y} = 256$ and $3^{x-y} = 729$.

   Change 256 to 2^8 and 729 to 3^6. (Always change to smallest base.) $2^{x+y} = 2^8$ and $3^{x-y} = 3^6$; set the exponents equal to create a system of equations.

   $x + y = 8$

   $x - y = 6$

   $2x = 14$

   $x = 7$. Only x is asked for.

3. Solve for x: $\left(\frac{1}{16}\right)^{x-3} = (32)^{x+3}$.

   Change to base 2. $(2^{-4})^{x-3} = (2^5)^{x+3}$.

   Multiply exponents and set them equal to each other.

   $-4(x - 3) = 5(x + 3) \Rightarrow -4x + 12 = 5x + 15 \Rightarrow 9x = -3$.

   $\therefore x = -\frac{1}{3}$.

4. Solve for x: $\sqrt{125^x} = \frac{5}{25^x}$.

   $\sqrt{(5^3)^x} = \frac{5}{(5^2)^x} \Rightarrow (5^{3x})^{\frac{1}{2}} = \frac{5}{5^{2x}}$; multiply both sides by 5^{2x}.

   $\therefore 5^{\frac{3x}{2}} \cdot 5^{2x} = 5^1$; add the exponents on the left side and form an equation.

   $5^{\frac{7x}{2}} = 5^1 \Rightarrow \frac{7x}{2} = 1 \Rightarrow 7x = 2, x = \frac{2}{7}$.

5. Find all real numbers x for which $\dfrac{3^{\sqrt{12x}} + 3}{4} = 3^{\sqrt{3x}}$.

Multiply both sides by 4 and simplify $\sqrt{12x}$ to $2\sqrt{3x}$.

$\therefore\ 3^{2\sqrt{3x}} + 3 = 4 \cdot 3^{\sqrt{3x}}$; rearrange terms to $3^{2\sqrt{3x}} - 4 \cdot 3^{\sqrt{3x}} + 3 = 0$.

$3^{2\sqrt{3x}} - 4 \cdot 3^{\sqrt{3x}} + 3 = 0$ is quadratic in form, so let $a = 3^{\sqrt{3x}}$, $a^2 - 4a = 3 = 0$.

Factor, $(a - 3)(a - 1) = 0 \Rightarrow a = 1, 3$. $\therefore$ Because $a = 3^{\sqrt{3x}}$, substitute for a.

$3^{\sqrt{3x}} = 3^1 \Rightarrow \sqrt{3x} = 1$; square both sides,

$3x = 1, x = \dfrac{1}{3}$, and $3^{\sqrt{3x}} = 1$; then $3^{\sqrt{3x}} = 3^0, \sqrt{3x} = 0, x = 0$.

$\therefore\ x = 0, \dfrac{1}{3}$.

6. Solve for x: $\sqrt{\dfrac{9^{x+3}}{27^x}} = 81$.

Change to smallest base, $\sqrt{\dfrac{9^{x+3}}{27^x}} = \sqrt{\dfrac{(3^2)^{x+3}}{3^{3x}}} = 3^4$.

Square both sides, $\left(\sqrt{\dfrac{(3^2)^{x+3}}{3^{3x}}}\right)^2 = (3^4)^2 \Rightarrow \dfrac{3^{2x+6}}{3^{3x}} = 3^8$.

Subtract exponents, $3^{2x+6-3x} = 3^8$; set exponents equal and simplify.

$\therefore\ -x + 6 = 8 \Rightarrow x = -2$.

7. Given: $5^x + 2^y = 2^x + 5^y = \dfrac{7}{10}$, find $(x + y)^{-1}$.

You need to find the two rational numbers that add up to $\dfrac{7}{10}$

and what exponents will work.

$\therefore\ \dfrac{5}{10} + \dfrac{2}{10} = \dfrac{7}{10}$ and reducing, $\dfrac{1}{2} + \dfrac{1}{5} = \dfrac{7}{10}$, and $\dfrac{1}{2} = 2^{-1}$ and

$\dfrac{1}{5} = 5^{-1}$, so $x = y = -1$.

$\therefore\ (x + y)^{-1} = (-1 - 1)^{-1} = (-2)^{-1} = -\dfrac{1}{2}$.

$\therefore$ The answer is $-\dfrac{1}{2}$.

8. Find x: $4^{x+1} - 5 \cdot 2^x + 1 = 0$.

   Change to base 2, $(2^2)^{x+1} = 2^{2x+2} = 2^{2x} \cdot 2^2$.

   $\therefore (2^x)^2 \cdot 2^2 - 5 \cdot 2^x + 1 = 0$; let $a = 2^x$, then $a^2 \cdot 4 - 5a + 1 = 0$.

   $4a^2 - 5a + 1 = 0$; factor, $(4a - 1)(a - 1) = 0 \Rightarrow a = \frac{1}{4}, 1$.

   Because $a = 2^x$, then setting exponents equal,

   $2^x = \frac{1}{4} = 2^{-2}$, so $x = -2$ and $2^x = 1 = 2^0$, so $x = 0$.

   $\therefore$ The answer is $x = 0, -2$.

Logarithmic Functions

1. Find all real x such that $\log_3(x + 3) = 1 - \log_3(x + 5)$.

   Rearrange terms, $\log_3(x + 3) + \log_3(x + 5) = 1$.

   $\log_3[(x + 3)(x + 5)] = 1$; use the log of a product rule.

   $(x + 3)(x + 5) = 3^1$; change to exponential form and expand,

   $x^2 + 8x + 12 = 0$; factor,

   $(x + 6)(x + 2) = 0$

   $x = -6, -2$; reject -6 because this leads to log(negative numbers).

   $\therefore$ The answer is $x = -2$.

2. Solve for x: $\log_2[\log_3(x + 1)] = -1$.

   Let $z = \log_3(x + 1)$.

   $\log_2 z = -1$; change to exponential form, $z = \frac{1}{2}$.

   $\log_3(x + 1) = \frac{1}{2} \Rightarrow x + 1 = 3^{\frac{1}{2}} = \sqrt{3}$.

   $\therefore x = \sqrt{3} - 1$.

3. What is the simplified numerical value of the sum

   $\log_2\left(1 - \frac{1}{2}\right) + \log_2\left(1 - \frac{1}{3}\right) + \ldots + \log_2\left(1 - \frac{1}{64}\right)$,

   where the nth term is $\log_2\left(1 - \frac{1}{n + 1}\right)$?

   $\log_2\left(1 - \frac{1}{2}\right) + \log_2\left(1 - \frac{1}{3}\right) + \ldots + \log_2\left(1 - \frac{1}{64}\right)$

$$= \log_2\left(\frac{1}{2}\cdot\frac{2}{3}\cdot\ldots\cdot\frac{62}{63}\cdot\frac{63}{64}\right) = \log_2\left(\frac{1}{64}\right).$$

The process is called telescoping as every term cancels except the first numerator and the last denominator.

Because $\log_2\left(\frac{1}{64}\right) = x$ (some number), change to exponential form.

$\therefore\ 2^x = 2^{-6},\ x = -6.$

4. If $\log_{10}2 = a$ and $\log_{10}3 = b$, find the exact solution to the equation $12^{x+2} = 18^{x-3}$ in terms of a and b.

All logs are base 10, so you don't need to write the base.

Take log base 10 of both sides and use the rule $\log_b a^n = n \cdot \log_b a$.

$12^{x+2} = 18^{x-3} \Rightarrow (x+2)\log 12 = (x-3)\log 18$; distribute

$x\log 12 + 2\log 12 = x\log 18 - 3\log 18$, combine the x terms,

$x\log 18 - x\log 12 = 2\log 12 + 3\log 18$; factor out an x for the left side.

$x(\log 18 - \log 12) = 2\log 12 + 3\log 18$; divide by $(\log 18 - \log 12)$ and

$$x = \frac{2\log 12 + 3\log 18}{\log 18 - \log 12}$$

$\dfrac{2\log 12 + 3\log 18}{\log 18 - \log 12} = \dfrac{2(2\log 2 + \log 3) + 3(2\log 3 + \log 2)}{(2\log 3 + \log 2) - (2\log 2 + \log 3)}$; use the rule for product of logs and rule for $\log_b a^n$.

In $\dfrac{2(2\log 2 + \log 3) + 3(2\log 3 + \log 2)}{(2\log 3 + \log 2) - (2\log 2 + \log 3)}$, replace the a's and b's from the problem.

$$x = \frac{2(2a+b) + 3(2b+a)}{(2b+a) - (2a+b)} = \frac{7a+8b}{b-a}.$$

5. If $\log_{10}x^n = 100x^2$ and $\log_x 10 = n$, find x.

Change to exponential form,

$10^{100x^2} = x^n$ and $x^n = 10$

$10^{100x^2} = 10^1$

$100x^2 = 1 \Rightarrow x^2 = \frac{1}{100}$

$x = \pm\frac{1}{10}$; reject $-\frac{1}{10}$ as $\log_{-\frac{1}{10}}10$ is undefined.

$\therefore\ x = \frac{1}{10}.$

6. Solve for x: $x^{\log_2 x} = 16x^3$.

   Take the log base 2 of both sides, $\log_2 x^{\log_2 x} = \log_2 16x^3$, and use $\log_b a^n$ rule,

   $\log_2 x \cdot \log_2 x \Rightarrow (\log_2 x)^2 = \log_2 16x^3$,

   and $\log_2 16x^3 = \log_2 16 + \log_2 x^3$; use log of a product rule,

   $\log_2 16 = 4$.

   $\therefore (\log_2 x)^2 = 4 + 3\log_2 x \Rightarrow$ let $a = \log_2 x$ and rearrange equation,

   $a^2 - 3a - 4 = 0 \Rightarrow (a-4)(a+1) = 0 \Rightarrow a = 4, -1$.

   Because $a = \log_2 x$, $\log_2 x = 4 \Rightarrow x = 16$, $\log_2 x = -1 \Rightarrow x = \frac{1}{2}$.

   $\therefore x = \frac{1}{2}, 16$.

   Alternate solution: Let $\log_2 x = a \Rightarrow x = 2^a$

   $(2^a)^a = 2^4(2^a)^3 \Rightarrow 2^{a^2} = 2^{4+3a} \Rightarrow a^2 - 3a - 4 = 0$, etc.

7. Solve: $\frac{625}{x^{\log_5 x}} = 1$.

   Cross-multiply, $625 = x^{\log_5 x}$; take log base 5 of both sides and $\log_5(625) = 4$.

   $\therefore 4 = \log_5 x^{\log_5 x}$, $4 = (\log_5 x)^2 \Rightarrow \log_5 x = \pm 2$; change to exponential form.

   $\therefore x = 5^2, x = 5^{-2} \Rightarrow x = 25, \frac{1}{25}$.

8. Let $r = \log_b\left(\frac{8}{45}\right)$ and $s = \log_b\left(\frac{135}{4}\right)$. Find an ordered pair of integers (m, n) such that $\log_b\left(\frac{32}{5}\right) = mr + ns$.

   $$\begin{aligned}\log_b(2^5 \cdot 5^{-1}) &= m\log_b(3^{-2} \cdot 5^{-1} \cdot 2^3) + n\log_b(2^{-2} \cdot 5 \cdot 3^3)\\ &= \log_b(3^{-2} \cdot 5^{-1} \cdot 2^3)^m + \log_b(2^{-2} \cdot 5 \cdot 3^3)^n.\end{aligned}$$

   Use the $\log_b a^n = n \cdot \log_b a$ rule.

   $\log_b(2^5 \cdot 5^{-1}) = \log_b(3^{-2m} \cdot 5^{-m} \cdot 2^{3m}) + \log_b(2^{-2n} \cdot 5^n \cdot 3^{3n})$.

   Use the $(ab)^n$ rule.

   $\log_b(2^5 \cdot 5^{-1}) = \log_b(2^{3m} \cdot 5^{-m} \cdot 3^{-2m} \cdot 3^{3n} \cdot 5^n \cdot 2^{-2n})$,

   using log ab rule.

Form a system of equations by setting bases equal to each other.

a. $2^5 = 2^{3m} \cdot 2^{-2n} \Rightarrow 5 = 3m - 2n$

b. $5^{-1} = 5^{-m} \cdot 5^n \Rightarrow -1 = -m + n$

c. $3^0 = 3^{-2m} \cdot 3^{3n} \Rightarrow 0 = -2m + 3n$

Solve by multiplying equation *a* by 2 and *c* by 3 and adding them together.

$10 = 6m - 4n$

$0 = -6m + 9n$

$10 = 5n \Rightarrow n = 2.$

Substituting in equation *b*,

$-1 = -m + 2 \Rightarrow m = 3.$

$\therefore$ The answer is $(3, 2)$.

9. If $(\log_3 x)(\log_x 2x)(\log_{2x} y) = \log_x x^2$, compute the numerical value of *y*.

Change to base *x* and use the

change of base formula, $\log_b a = \dfrac{\log_c a}{\log_c b}$

$\dfrac{\log_x x}{\log_x 3} \cdot \dfrac{\log_x 2x}{\log_x x} \cdot \dfrac{\log_x y}{\log_x 2x} = 2 \log_x x$; use $\log_b a^n$ rule,

cancel like terms, and $\log_x x = 1$.

$\dfrac{\log_x y}{\log_x 3} = 2 \Rightarrow \log_x y = 2 \log_x 3.$

$\log_x y = \log_x 3^2 \Rightarrow$ use if $\log_b x_1 = \log_b x_2$, then $x_1 = x_2$.

$\therefore y = 3^2 = 9.$

10. Given $\log_{18} 6 = a$, find $\log_{18} 16$ in terms of *a*.

$\log_{18} 6 = \log_{18}(2 \cdot 3) = \log_{18} 2 + \log_{18} 3;$

log of a product rule.

Clever step to introduce $\log_{18} 3$, $\log_{18} 3 = \log_{18}(18 \div 6)$

$\Rightarrow \log_{18} 18 - \log_{18} 6$; $\log \frac{a}{b}$ rule.

$\log_{18} 18 = 1$ and $\log_{18} 6 = a$ (from the given).

Substituting both, we get $\log_{18} 3 = (1 - a)$.

$\therefore$ Using information from above, $\log_{18} 6 = \log_{18} 2 + (1 - a)$

$\Rightarrow a = \log_{18} 2 + (1 - a)$ and $\log_{18} 2 = 2a - 1$.

$\therefore \log_{18} 16 = \log_{18} 2^4 = 4 \log_{18} 2 = 4(2a - 1) = 8a - 4.$

Solving Systems of Equations

1. Solve for the ordered pair (x, y).

$$\frac{6}{x} + \frac{5}{y} = 1$$

$$\frac{3}{x} - \frac{10}{y} = 3$$

Let $a = \frac{1}{x}$ and $b = \frac{1}{y}$, and multiply the first equation by 2 and then add it to the other equation.

$$\therefore 6a + 5b = 1 \Rightarrow 12a + 10b = 2$$

$$3a - 10b = 3$$

$$15a = 5 \Rightarrow a = \frac{1}{3}$$

Substitute $a = \frac{1}{3}$ in the equation $6a + 5b = 1$

$$6\left(\frac{1}{3}\right) + 5b = 1 \Rightarrow 2 + 5b = 1$$

$5b = -1 \Rightarrow b = -\frac{1}{5}$, and because $b = \frac{1}{y}$, $y = -5$

and because $a = \frac{1}{x}$, $x = 3$.

$\therefore (x, y) = (3, -5)$.

2. Find all ordered pairs (t, w) that make the following statements true:

$\frac{2}{t} = 7 - \frac{3}{2w}$ and $\frac{6}{w} + \frac{1}{t} = 0$.

$\frac{1}{t} = -\frac{6}{w} \Rightarrow$ Substitute in the other equation,

$$2\left(-\frac{6}{w}\right) = 7 - \frac{3}{2w} \Rightarrow -\frac{12}{w} + \frac{3}{2w} = 7.$$

Multiply by $2w$, $-24 + 3 = 14w \Rightarrow -21 = 14w \Rightarrow w = -\frac{3}{2}$.

Because $\frac{6}{w} = 6 \cdot \frac{1}{w}$ and if $w = -\frac{3}{2}$, then

$$6\left(-\frac{2}{3}\right) + \frac{1}{t} = 0 \Rightarrow -4 = -\frac{1}{t} \Rightarrow t = \frac{1}{4}.$$

$\therefore (t, w) = \left(\frac{1}{4}, -\frac{3}{2}\right)$.

3. Solve for the ordered pair (x, y).

   $2x + \sqrt{5}y = 7$ and $\sqrt{5}x - 3y = -2\sqrt{5}$.

   Multiply the first equation by $\sqrt{5}$ and the second one by -2 and then add them.

   $\sqrt{5}(2x + \sqrt{5}y = 7)$

   $-2(\sqrt{5}x - 3y = -2\sqrt{5})$

   $2\sqrt{5}x + 5y = 7\sqrt{5}$

   $-2\sqrt{5}x + 6y = 4\sqrt{5}$

   $11y = 11\sqrt{5} \Rightarrow y = \sqrt{5}$; substitute the value of y in the first equation.

   $2x + \sqrt{5} \cdot \sqrt{5} = 7 \Rightarrow 2x + 5 = 7 \Rightarrow x = 1$.

   $\therefore (x, y) = (1, \sqrt{5})$.

4. Find all ordered pairs (x, y) in terms of a that satisfy the following system of equations:

   $3ax + 4ay = 1$ and $\frac{x}{3} - \frac{y}{2} = \frac{2}{a}$.

   Multiply the first equation by $\frac{3}{a}$ and the second by 24.

   There are other numbers to use but this simplifies the problem.

   $9x + 12y = \frac{3}{a}$

   $8x - 12y = \frac{48}{a}$

   Add the two equations.

   $17x = \frac{51}{a} \Rightarrow x = \frac{3}{a}$; substitute the value of x in the first equation.

   $3a\left(\frac{3}{a}\right) + 4ay = 1 \Rightarrow 9 + 4ay = 1,\ 4ay = -8 \Rightarrow y = -\frac{2}{a}$,

   $\therefore (x, y) = \left(\frac{3}{a}, -\frac{2}{a}\right)$.

5. The graphs of $y = -x^2 + x - 1$ and $\frac{(y+3)}{(x-2)} = 2$ have only one point of intersection. Find this ordered pair (x, y).

   Multiply the second equation by $x - 2$ and simplify, $y = 2(x - 2) - 3 \Rightarrow y = 2x - 7$.

   Set the two equations equal as they both equal y; $2x - 7 = -x^2 + x - 1$;

   combine similar terms, $x^2 + x - 6 = 0$ and factor,

   $(x + 3)(x - 2) = 0 \Rightarrow x = -3, 2$ but substituting 2 in the denominator makes it equal to 0, so reject it and the only answer for x is -3.

Substituting in the first equation, $y = -(-3)^2 - 3 - 1 = -13$.

$\therefore$ The answer is $(-3, -13)$.

6. Find the value of $x^2 + y^2$ if (x, y) is a solution of the system of equations: $xy = 5$ and $x^2y + 2x = xy^2 + 2y + 35$.

   Rearrange the terms in the second equation, $x^2y - xy^2 + 2x - 2y = 35$;

   factor, $xy(x-y) + 2(x-y) = 35 \Rightarrow (x-y)(xy+2) = 35$.

   Substitute $xy = 5$ in for $xy + 2$ and get 7.

   $\therefore$ $7(x-y) = 35 \Rightarrow x - y = 5$; now square this expression

   and get $x^2 - 2xy + y^2 = 25$, $-2xy = -10$ because $xy = 5$; substitute

   and $x^2 - 10 + y^2 = 25$.

   $\therefore$ The answer is $x^2 + y^2 = 35$.

   Again, we didn't find x and y but the value of $x^2 + y^2$.

   This means you need to keep the $x^2 + y^2$ expression in the problem.

7. Find the intersection points (in simplest radical form) of the circle $x^2 + y^2 = 40$ and $x^2 + 2xy - 3y^2 = 0$.

   Factor $x^2 - 2xy - 3y^2 = 0$, $(x+3y)(x-y) = 0$. $\therefore$ $x = -3y$, $x = y$;

   substitute $x = -3y$, $(-3y)^2 + y^2 = 40 \Rightarrow 10y^2 = 40$, $y = \pm 2$.

   If $x = -3y$ and $y = \pm 2$, we get $(-6, 2)$ and $(6, -2)$.

   Also, because $x = y$, $y^2 + y^2 = 40 \Rightarrow y^2 = 20$, $y = \pm 2\sqrt{5}$.

   As $x = y$, we get $(2\sqrt{5}, 2\sqrt{5})$ and $(-2\sqrt{5}, -2\sqrt{5})$.

   $\therefore$ There are four points of intersection, and the answers are

   $(-6, 2)$, $(6, -2)$, $(2\sqrt{5}, 2\sqrt{5})$, $(-2\sqrt{5}, -2\sqrt{5})$.

8. If $abc > 0$, find the ordered triple that satisfies $ab = 24$, $ac = 72$, and $bc = 108$.

   Many students would try to solve this problem by substitution. The following is a nice solution.

   Divide $ac = 72$ by $ab = 24$. $\therefore$ $\dfrac{ac}{ab} = \dfrac{72}{24} \Rightarrow \dfrac{c}{b} = 3$, then $c = 3b$.

   Substitute this for $bc = 108$. $\therefore$ $(b)(3b) = 3b^2 = 108$, $b^2 = 36$,

   then $b = 6$ as $b > 0$. Because $ab = 24$, then $a = 4$ and $6c = 108$ so $c = 18$.

   $\therefore$ The answer is $(4, 6, 18)$.

Matrices and Determinants

1. Find the value of x: $\begin{vmatrix} 2 & 3 \\ 7 & 1 \end{vmatrix} = \begin{vmatrix} 4 & 6 \\ 14 & x \end{vmatrix}$.

   Use the definition for evaluating a determinant, $\begin{vmatrix} a & b \\ c & d \end{vmatrix} = ad - cb$.

   $2 \cdot 1 - 7 \cdot 3 = -19$

   $4x - 14 \cdot 6 = 4x - 84$

   $-19 = 4x - 84$

   $4x = 65$

   $\therefore$ The answer is $x = \frac{65}{4}$.

2. Find all values of x for which $\dfrac{\begin{vmatrix} x & 2 \\ 2 & x \end{vmatrix}}{\begin{vmatrix} x & x \\ x & 2 \end{vmatrix}} = -\frac{5}{3}$.

   $\frac{x^2 - 4}{2x - x^2} = -\frac{5}{3}$; cross-multiply,

   $3x^2 - 12 = -10x + 5x^2$

   $2x^2 - 10x + 12 = 0$; divide by 2,

   $x^2 - 5x + 6 = 0$; factor,

   $(x - 2)(x - 3) = 0$.

   $x = 2, 3$ but $x = 2$ makes the denominator equal to 0, so reject it.

   $\therefore$ The answer is $x = 3$.

3. Solve for x and y. Express your answer as an ordered pair (x, y).

   $\begin{vmatrix} x-3 & y+1 \\ 17 & -3 \end{vmatrix} = -3$ and $\begin{vmatrix} y-2 & x+4 \\ 2 & -19 \end{vmatrix} = 18$.

   Create a system of equations by evaluating the left determinant,

   $-3(x - 3) - 17(y + 1) = -3$

   $-3x + 9 - 17y - 17 = -3$

   $^*-3x - 17y = 5$

   Do the same for the determinant on the right.

   $-19(y - 2) - 2(x + 4) = 18$

$-19y + 38 - 2x - 8 = 18$

$^{*}-2x - 19y = -12$

Multiply so you can eliminate the x's,

$-3(-2x - 19y = -12)$

$^{**}6x + 57y = 36$

$2(-3x - 17y = 5)$

$^{**}-6x - 34y = 10$

Solving the ** equations,

$23y = 46 \Rightarrow y = 2;$

substituting in, $-3x - 34 = 5$

$-3x = 39 \Rightarrow x = -13.$

$\therefore\ (x, y) = (-13, 2).$

4. Evaluate the determinant $\begin{vmatrix} 0 & 2 & -3 \\ 3 & 5 & -3 \\ 1 & 2 & 0 \end{vmatrix}$.

Repeat the first two columns and use the rule for evaluating a 3×3 determinant.

$$\begin{matrix} 0 & 2 & -3 & 0 & 2 \\ 3 & 5 & -3 & 3 & 5 \\ 1 & 2 & 0 & 1 & 2 \end{matrix}$$

$0 + (-6) - 18 - (-15 + 0 + 0)$

$= -24 + 15$

$\therefore$ The answer is -9.

5. Solve for x: $\begin{vmatrix} x^2 & 3x & 5 \\ 1 & 3 & 5 \\ 4 & -6 & 5 \end{vmatrix} = 0.$

Find the value of the determinant as you did on problem 4:

$$\begin{vmatrix} x^2 & 3x & 5 \\ 1 & 3 & 5 \\ 4 & -6 & 5 \end{vmatrix} = \begin{matrix} x^2 & 3x & 5 & x^2 & 3x \\ 1 & 3 & 5 & 1 & 3 \\ 4 & -6 & 5 & 4 & -6 \end{matrix} =$$

$15x^2 + 60x - 30 - (60 - 30x^2 + 15x) = 0$

$15x^2 + 60x - 30 - 60 + 30x^2 - 15x = 0$

$45x^2 + 45x - 90 = 0 \Rightarrow x^2 + x - 2 = 0$

$(x + 2)(x - 1) = 0$

$\therefore$ The answer is $x = -2, 1$.

6. If $A = \begin{bmatrix} 1 & 3 \\ 4 & 3 \end{bmatrix}$ and $B = \begin{bmatrix} x+1 \\ y \end{bmatrix}$, find x and y such that $A \cdot B = 3 \cdot B$.

$A \times B$

$\begin{bmatrix} 1 & 3 \\ 4 & 3 \end{bmatrix} \cdot \begin{bmatrix} x+1 \\ y \end{bmatrix} = \begin{bmatrix} x+1+3y \\ 4x+4+3y \end{bmatrix}$ and $3\begin{bmatrix} x+1 \\ y \end{bmatrix} = \begin{bmatrix} 3x+3 \\ 3y \end{bmatrix}$.

3 is a scalar, so multiply each element by 3.

$\therefore \begin{bmatrix} x+1+3y \\ 4x+4+3y \end{bmatrix} = \begin{bmatrix} 3x+3 \\ 3y \end{bmatrix}$; two matrices are equal if they are the same size—both are 2×1—and if the elements are the same.

$\therefore$ Set the rows of A and B equal.

$x + 1 + 3y = 3x + 3 \Rightarrow 3y = 2x + 2$ and

$4x + 4 + 3y = 3y \Rightarrow 4x = -4, x = -1$

Because $x = -1$ and $3y = 2x + 2$, substitute $x = -1$, $3y = -2 + 2 = 0$, so $y = 0$.

$\therefore$ $x = -1$ and $y = 0$.

7. Compute $3\begin{bmatrix} 2 & -5 & 1 \\ 3 & 0 & -4 \end{bmatrix} - 2\begin{bmatrix} 1 & -2 & -3 \\ 1 & -1 & 4 \end{bmatrix}$.

Multiply both matrices by the scalars 3 and 2, respectively.

$\therefore \begin{bmatrix} 6 & -15 & 3 \\ 9 & 0 & -12 \end{bmatrix} - \begin{bmatrix} 2 & -4 & -6 \\ 2 & -2 & 8 \end{bmatrix}$;

now subtract the two matrices as they have the same dimensions.

$\therefore$ The answer is $\begin{bmatrix} 4 & -11 & 9 \\ 7 & 2 & -20 \end{bmatrix}$.

8. Find the ordered pair (x, y) such that $\begin{bmatrix} x & y \\ 2 & 4 \end{bmatrix} \cdot \begin{bmatrix} 1 & y-4 \\ 2 & x \end{bmatrix} = \begin{bmatrix} -1 & -50 \\ 10 & 6 \end{bmatrix}$.

   Multiply the matrices:
   first row by first column and set equal to -1, $x + 2y = -1$, and
   second row by second column and set equal to 6, $2(y-4) + 4x = 6 \Rightarrow$
   $4x + 2y = 14$; subtract the first equation,
   $-x - 2y = 1$
   $3x = 15$ and $x = 5$; substituting, $5 + 2y = -1$, $y = -3$.
   $\therefore$ The answer is $(x, y) = (5, -3)$.

Sequences and Series

1. If $\frac{1}{3}$, $\frac{1}{x}$, and $\frac{1}{7}$ form an arithmetic progression in that order, compute x.

   Because the problem states "in that order," $\frac{1}{x}$ is the arithmetic mean (average) of

   $\frac{1}{3}$ and $\frac{1}{7}$. $\therefore$ $\frac{1}{x} = \frac{\frac{1}{3}+\frac{1}{7}}{2} \Rightarrow \frac{1}{x} = \frac{\frac{10}{21}}{2}$; multiply by $\frac{\frac{1}{2}}{\frac{1}{2}}$; then $\frac{1}{x} = \frac{5}{21}$.

   $\therefore$ The answer is $x = \frac{21}{5}$.

2. Find $\sum_{n=1}^{12}\left(\frac{1}{n+2} - \frac{1}{n+3}\right)$.

   Replacing the numbers 1 to 12 for n and grouping,
   $\left(\frac{1}{3}-\frac{1}{4}\right)+\left(\frac{1}{4}-\frac{1}{5}\right)+\left(\frac{1}{5}-\frac{1}{6}\right)+\ldots+\left(\frac{1}{14}-\frac{1}{15}\right)$, you see that all terms cancel except the

   first and last terms, so $\frac{1}{3} - \frac{1}{15} = \frac{4}{15}$.

   $\therefore$ The answer is $\frac{4}{15}$.

3. The smallest interior angle of a polygon with 54 diagonals measures 132°. If the measures of all the interior angles are in arithmetic progression (A.P.), find the measure of the largest angle of the polygon.

   Use the formula for the number of diagonals in a polygon, $\frac{n(n-3)}{2} = 54$.

$\therefore n^2 - 3n - 108 = 0$; factor $(n-12)(n+9) = 0 \Rightarrow n = 12$; reject -9 as n represents the number of sides.

$\therefore n = 12$ and the sum of all the interior angles in a polygon is

$180(n-2) = 180(12-2) = 180 \cdot 10 = 1{,}800$.

$\therefore$ Use the formula for the sum of the terms in an A.P.,

$\therefore S_{12} = \frac{12}{2}(2 \cdot 132 + 11d) = 1{,}800 \Rightarrow 6(264 + 11d) = 1{,}800$

$264 + 11d = 300$ (dividing by 6).

$\therefore 11d = 36 \Rightarrow d = \frac{36}{11}$. Use the formula for the last term of an A.P. (this will be the largest angle), $\therefore a_{12} = 132 + 11d$ and because $d = \frac{36}{11}$,

$a_{12} = 132 + 11 \cdot \frac{36}{11} \Rightarrow 132 + 36 = 168$.

$\therefore$ The largest angle is 168°.

4. Find the first term of an infinite geometric progression for which the sum is $2\sqrt{2} + 2$ and for which the common ratio is $\frac{1}{\sqrt{2}}$.

   Use the formula $S = \frac{a_1}{1-r}$, where a_1 is the first term. $\therefore 2\sqrt{2} + 2 = \frac{a_1}{1 - \frac{1}{\sqrt{2}}}$.

   Simplify the denominator, $1 - \frac{1}{\sqrt{2}} = \frac{\sqrt{2}-1}{\sqrt{2}}$; replace in problem, $\frac{a}{\frac{\sqrt{2}-1}{\sqrt{2}}}$; multiply by $\frac{\sqrt{2}}{\sqrt{2}}$ and get $\frac{\sqrt{2}a}{\sqrt{2}-1}$.

   $\therefore 2\sqrt{2} + 2 = \frac{\sqrt{2}a}{\sqrt{2}-1} \Rightarrow 2(\sqrt{2}+1)(\sqrt{2}-1) = \sqrt{2}a$

   $2 = \sqrt{2}a \Rightarrow a = \frac{2}{\sqrt{2}}$; rationalize, $\frac{2}{\sqrt{2}} \cdot \frac{\sqrt{2}}{\sqrt{2}} = \frac{2\sqrt{2}}{2} = \sqrt{2}$.

   $\therefore$ The answer is $a = \sqrt{2}$.

5. Find all possible values of X and Y such that 3, X, and Y will be in an arithmetic progression and X, Y, and 8 will be in a geometric progression. Express the answer(s) as ordered pair(s) (X, Y).

   3, X, and Y are in an A.P. $\therefore X - 3 = Y - X \Rightarrow Y = 2X - 3$

X, Y, and 8 are a G.P., $\frac{Y}{X} = \frac{8}{Y}$ and solving for Y, $Y^2 = 8X \Rightarrow Y = \pm\sqrt{8X}$; setting the Y's equal, $\pm\sqrt{8X} = 2X - 3$; square both sides, $8X = 4X^2 - 12X + 9$; combine, $4X^2 - 20X + 9 = 0$; factor, $(2X - 9)(2X - 1) = 0 \Rightarrow X = \frac{9}{2}, \frac{1}{2}$; substitute in $Y = 2X - 3$; if $X = \frac{9}{2}$, $Y = 6$ and if $X = \frac{1}{2}$, $Y = -2$.

$\therefore$ The answers are $\left(\frac{9}{2}, 6\right)$ and $\left(\frac{1}{2}, -2\right)$.

6. The sum of n positive terms of an arithmetic series is 216. The first term is n and the last term is twice the first term.
   What is the common difference?

   $216 = \frac{n}{2}(n + 2n) \Rightarrow 432 = 3n^2$, $n^2 = 144 \Rightarrow n = \pm 12$.

   Because $n > 0$, $n = 12$. $\therefore$ $24 = 12 + 11d$.

   $\therefore$ The answer is $d = \frac{12}{11}$.

7. An infinite geometric progression has a first term 1, and the sum of all terms is $\frac{9}{2}$ times its second term. What possible values could the common ratio have?
   The sequence is $1, a, a^2, a^3, \ldots$

   The sum is $\frac{1}{(1-a)}$, which equals $\frac{9}{2}a$, cross-multiplying, we get $2 = 9a - 9a^2$; rearrange and factor, $9a^2 - 9a + 2 = 0 \Rightarrow (3a - 1)(3a - 2) = 0$.

   $\therefore$ The answer is $a = \frac{1}{3}, \frac{2}{3}$.

8. In a geometric series of positive terms, the difference between the fifth and fourth term is 243, and the difference between the second and first term is 9. What is the sum of the first five terms of the series?

   $S_5 = a + ar + ar^2 + ar^3 + ar^4 = \frac{a(1 - r^5)}{1 - r}$.

   $ar^4 - ar^3 = 243$ and $ar - a = 9$. Divide the first equation by the second and get

   $\frac{ar^4 - ar^3}{ar - a} = \frac{243}{9} \Rightarrow \frac{ar^3(r-1)}{a(r-1)} = 27$.

   Cancel terms and take the cube root, so $r = 3$. Because $ar - a = 9$, $2a = 9 \Rightarrow a = \frac{9}{2}$.

Substituting in, $\dfrac{a(1-r^5)}{1-r} = \dfrac{\frac{9}{2}(1-3^5)}{1-3} = \dfrac{9}{2}\left(\dfrac{-242}{-2}\right) = \dfrac{1{,}089}{2}$.

$\therefore$ The sum of the first five terms is $544\ \frac{1}{2}$.

9. The arithmetic mean of two positive numbers exceeds their geometric mean by 50. By how much does the square root of the larger exceed the square root of the smaller?

   Let the numbers be a and b, $a > b$, then $\dfrac{a+b}{2} - \sqrt{ab} = 50$; multiply by 2,

   $a + b - 2\sqrt{ab} = 100$; rearrange to $a - 2\sqrt{ab} + b = 100$; factor, $(\sqrt{a} - \sqrt{b})^2 = 100$.

   Take the square root of both sides, $\sqrt{a} - \sqrt{b} = \pm 10$; as a and $b > 0$, so -10 is not a solution.

   $\therefore$ The answer is 10.

   Discussion. Notice again, we didn't find a or b. Always keep the question or answer form in mind. The factors of $a^2 - 2ab + b^2$ are $(a - b)^2$ so the factors of $a + b - 2\sqrt{ab}$ or $a - 2\sqrt{ab} + b$ are $(\sqrt{a} - \sqrt{b})^2$.

Analytical Geometry

1. Find the equation of a line, with integral slope, passing through (2, 2), that makes a triangle with an area 9 units^2 with the coordinate axes. Express the answer in the form of $Ax + By = C$.

   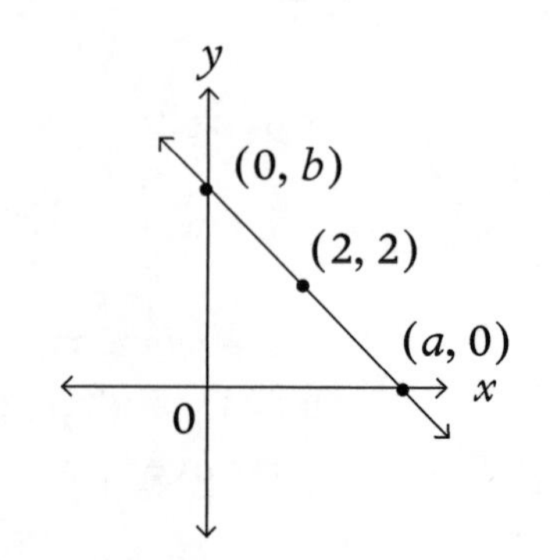

   Make a diagram and label the intercepts as $(0, b)$ and $(a, 0)$, draw a line through them and the point (2, 2). Use the intercept form for a linear equation: $\frac{x}{a} + \frac{y}{b} = 1$. We are going to set up a system of equations using the given information. Substitute $\frac{x}{a} + \frac{y}{b} = 1$, the ordered pair (2, 2) $\therefore\ \frac{2}{a} + \frac{2}{b} = 1$; multiply by ab, simplify and save this equation, $2b + 2a = ab$. Because the area of the triangle is 9, we can use $A = \frac{1}{2}ab$, where ab is the product of the x and y intercepts. $\therefore\ 9 = \frac{1}{2}ab \Rightarrow ab = 18$ and $a = \frac{18}{b}$; substitute this into the equation you saved. $2b + 2\left(\frac{18}{b}\right) = 18$; multiply by b,

$2b^2 + 36 = 18b$; rearrange and divide by 2, $b^2 - 9b + 18 = 0$; factor, $(b-3)(b-6) = 0$; and $b = 3, 6$. Because $ab = 18$, then if $b = 3$, $a = 6$, and if $b = 6$, $a = 3$. You get two equations; $\frac{x}{6} + \frac{y}{3} = 1$ and $\frac{x}{3} + \frac{y}{6} = 1$. Simplifying both, we get $x + 2y = 6$ and $2x + y = 6$.

The former does not have integral slope, so the only answer is $2x + y = 6$.

$\therefore$ The answer in the form of $Ax + By = C$ is $2x + y = 6$.

2. Find the equation of the ellipse having x-intercepts $(6, 0)$ and $(-6, 0)$ and foci $(3, 0)$ and $(-3, 0)$. Write your answer in the form of $\frac{x^2}{a^2} + \frac{y^2}{b^2} = 1$.

   The x-axis is the major axis, so $a = 6$. $\therefore \frac{x}{6^2} + \frac{y}{b^2} = 1$.

   To find b^2, use $b^2 = a^2 - c^2$. $\therefore b^2 = 36 - 9 = 27$ (don't find b).

   $\therefore$ The equation is $\frac{x^2}{36} + \frac{y^2}{27} = 1$.

3. Given a parabola in the form of $y = ax^2 + bx + c$ that passes through the point $(4, 3)$ and has a vertex $V(6, -3)$, find the coefficient of the x^2 term (which is called the leading coefficient).

   The standard form of a parabola is $y = a(x-h)^2 + k$, where (h, k) is the vertex.

   $\therefore$ Substitute $(6, -3)$ and obtain $y = a(x-6)^2 - 3$. Because the parabola goes through the point $(4, 3)$, substitute the ordered pair, $3 = a(4-6)^2 - 3 \Rightarrow 3 = 4a - 3$.

   $6 = 4a \Rightarrow a = \frac{6}{4} = \frac{3}{2}$.

   $\therefore$ The coefficient of x^2 is $\frac{3}{2}$.

4. Find the distance between the two lines $3x + 4y = 12$ and $3x + 4y = 15$.

   Use the formula for finding the distance from a point to a line, $\frac{|Ax + By + C|}{\sqrt{A^2 + B^2}}$; however, we need to find a point. Select any ordered pair that lies on one of the lines. This becomes the (x, y) in the formula.

   The A, B, and C will be used from the other equation.

   Select $(0, 3)$ from the first equation and rewrite the second equation as $3x + 4y - 15 = 0$, as it must be in this form.

$$\therefore \frac{|Ax + By + C|}{\sqrt{A^2 + B^2}} = \frac{|3 \cdot 0 + 4 \cdot 3 - 15|}{\sqrt{3^2 + 4^2}} = \frac{|-3|}{\sqrt{25}} = \frac{3}{5}.$$

$\therefore$ The distance is $\frac{3}{5}$.

5. Find the distance between the centers of the two circles $x^2 + 4x + y^2 - 6y = 2$ and $x^2 + y^2 + 8y = 7$.
   Complete the square for both circles to determine their centers,
   $x^2 + 4x + y^2 - 6y = 2 \Rightarrow x^2 + 4x + 4 + y^2 - 6y + 9 = 2 + 4 + 9 = 15.$
   $\therefore$ Factor, $(x + 2)^2 + (y - 3)^2 = 15$, so the center is $(-2, 3)$.
   Repeating the process for $x^2 + y^2 + 8y = 7$, $(x - 0)^2 + y^2 + 8y + 16 = 7 + 16 = 23$;
   factor and obtain $(x - 0)^2 + (y + 4)^2 = 23$ and the center is $(0, -4)$.
   Use the distance formula for finding the distance between points or, in this example, the distance between the centers of the circles.
   $D = \sqrt{(x_1 - x_2)^2 + (y_1 - y_2)^2} = \sqrt{(0 + 2)^2 + (3 + 4)^2} = \sqrt{4 + 49} = \sqrt{53}.$
   $\therefore$ The answer is $\sqrt{53}$.

6. Given the ellipse $9x^2 + 16y^2 = 144$, find the length of the segment joining a vertex to the nearest focus.
   Change the ellipse to standard form by dividing all terms by 144.
   $\frac{9x^2}{144} + \frac{16y^2}{144} = 1 \Rightarrow \frac{x^2}{16} + \frac{y^2}{9} = 1$; the ellipse is centered around the origin and the length of the semi-major axis a ($\frac{1}{2}$ of the major axis) is $\sqrt{16} = 4$.
   The length of the semi-minor axis b is $\sqrt{9} = 3$. Using the formula $c = \sqrt{a^2 - b^2}$,
   $\therefore c = \sqrt{16 - 9} = \sqrt{7}.$
   $\therefore$ The length of the segment from a vertex to the nearest focus is $4 - \sqrt{7}$.

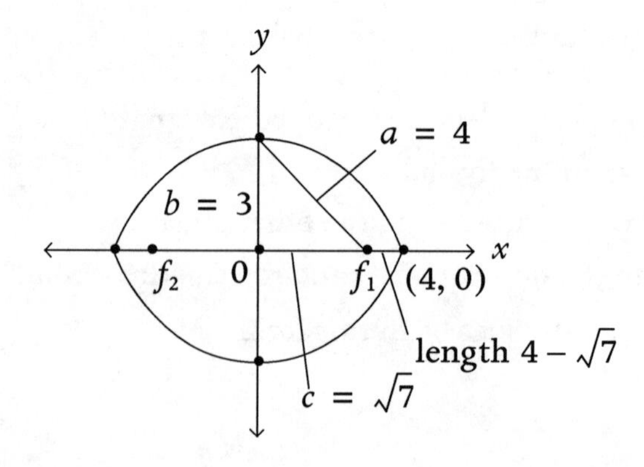

7. Find the equations of the asymptotes of the hyperbola for which the equation is $25y^2 - 16x^2 = 100$. Express your answer in the $y = mx + b$ form.
   Change to standard form by dividing all terms by 100.
   $$\frac{25y^2}{100} - \frac{16x^2}{100} = 1, \frac{y^2}{4} - \frac{4x^2}{25} = 1 \Rightarrow, \frac{y^2}{4} - \frac{x^2}{\frac{25}{4}} = 1.$$
   We need to find two points that the asymptotes pass through;
   one of them is the origin $(0, 0)$ and the other is $(\sqrt{b^2}, \sqrt{a^2})$,
   which equals $\left(\frac{5}{2}, 2\right)$. The slope is $\frac{2}{\frac{5}{2}}$, which equals $\frac{4}{5}$. Since the asymptotes pass through the orgin, their equations are of the form $y = \pm mx$.
   $\therefore$ The equations are $y = \pm\frac{4}{5}x$.

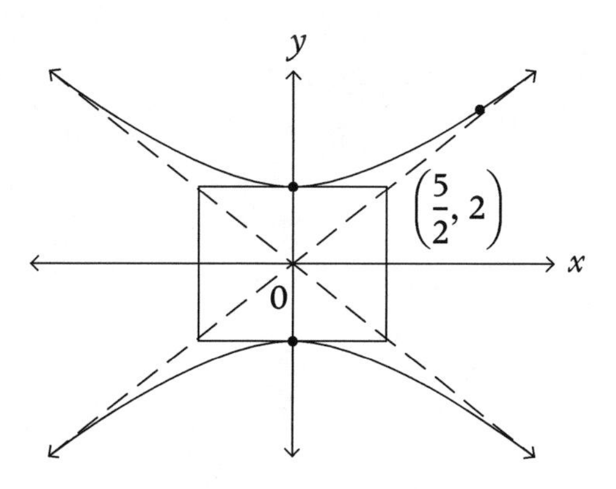

8. The ellipse $25(x-2)^2 + 4(y+3)^2 = 100$ is inscribed in a rectangle, the sides of which are parallel to the coordinate axes. What is the area of the rectangle in square units?
   Change the equation to standard form.
   $$\frac{25(x-2)^2}{100} + \frac{4(y+3)^2}{100} = 1$$
   $\Rightarrow \frac{(x-2)^2}{4} + \frac{(y+3)^2}{25} = 1$. The length of the major and minor axis are 4 and 10, respectively.
   $\therefore$ The area of the rectangle is 40. You should always draw a diagram to assist you in solving the problem.

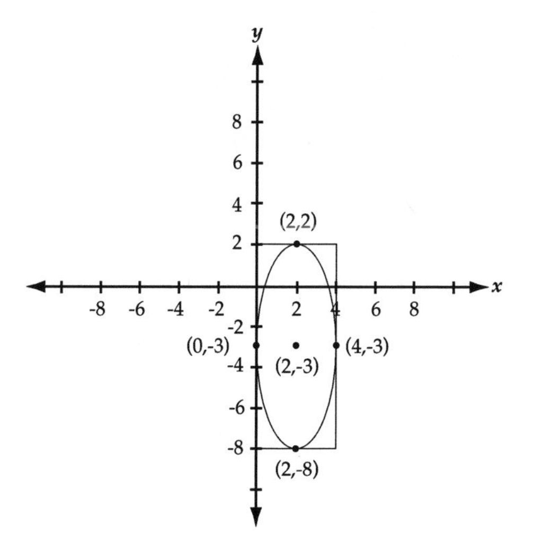

arithmetic mean (or average): commonly called the average, it is one-half the sum of two numbers

arithmetic means: the terms between two given terms of an arithmetic sequence

arithmetic sequence or progression: a sequence, such as the even numbers 2, 4, 6, 8, . . ., in which each term after the first is obtained by adding a constant, called the common difference, to the preceding term

arithmetic series: the sum of an arithmetic sequence or progression

complex number: a number of the form $a + bi$, where a and b are real numbers and i is a pure imaginary number; $i = \sqrt{-1}$

composition of functions: the composition of a function f with a function g is written $(f \circ g)(x) = f(g(x))$

degree of polynomial: the highest of the degrees of its terms

depressed equation or reduced equation: r is a root of the polynomial equation, $P(x) = 0$, and when synthetic division is performed, the remaining equation is called the depressed or reduced equation

determinant: a square array of numerals set off with vertical bars, which names a real number; the numerals in the array are called the entries.

dimensions of a matrix: the number of rows and the number of columns of entries in the matrix

discriminant: the number, $b^2 - 4ac$, inside the radical sign in the quadratic formula

double or multiple root: an equation that has roots repeated two or more times

exponential equation: an equation that has variable exponents

exponential function: a function defined by an equation of the form $y = b^x$, $b \neq 1$, and $b > 0$

finite sequence: a sequence that has a last term

function: a set of ordered pairs in which no first coordinate appears in more than one ordered pair; the set of first coordinates is called the domain, and the set of second coordinates is called the range of the function

geometric sequence or progression: a sequence, such as 1, 3, 9, 27, in which each term is multiplied by the same number to obtain the following terms; this number is called the common ratio.

geometric series: a series of which the terms are in geometric progression

inverse function: two functions f and g are inverses of each other if they are related such that $f(g(x)) = x$ for all x in the domain of g and $g(f(x)) = x$ for all x in the domain of f

leading coefficient: the coefficient of the term of highest degree in a polynomial

logarithm: in the exponential function, $a = b^y$, with base b, where $b > 0$ and $b \neq 1$, the exponent y is called the logarithm of a to the base b

matrix: a rectangular array of numbers. Each number in the array is called an entry of the matrix; the number of rows and the number of columns of entries in the matrix are its dimensions.

polynomial function: a function f is defined as $f(x) =$ $a_0x^n + a_1x^{n-1} + a_2x^{n-1} + \ldots + a_{n-1}x + a_n$, where n is a nonnegative integer and $a_0, a_1, \ldots, a_n$ are real numbers

quadratic function: a function f of the form $f(x) = ax^2 + bx + c$, where a, b, and c are real numbers and $a \neq 0$

rational algebraic expression: the quotient of two polynomials

sequence: a set of numbers in a particular order such as a, a^2, a^3, a^4

series: the sum of a sequence, $s_n = a + a^2 + a^3 + a^4 + \ldots + a^n$

zero of a function: any value of x for $f(x)$ that satisfies the equation $f(x) = 0$

Share Your Bright Ideas with Us!

We want to hear from you! Your valuable comments and suggestions will help us meet your current and future classroom needs.

Your name____Date________________

School name____

School address____

City ______________________State _____Zip__________Phone number (_________)_____________

Grade level taught________Subject area(s) taught________________________Average class size______

Where did you purchase this publication?____

Was your salesperson knowledgeable about this product? Yes_____ No_____

What monies were used to purchase this product?

___School supplemental budget ___Federal/state funding ___Personal

Please "grade" this Walch publication according to the following criteria:

Quality of service you received when purchasing	A	B	C	D	F
Ease of use	A	B	C	D	F
Quality of content	A	B	C	D	F
Page layout	A	B	C	D	F
Organization of material	A	B	C	D	F
Suitability for grade level	A	B	C	D	F
Instructional value	A	B	C	D	F

COMMENTS:____

____

What specific supplemental materials would help you meet your current—or future—instructional needs?

____

Have you used other Walch publications? If so, which ones?_________________________________

May we use your comments in upcoming communications? ___Yes ___No

Please **FAX** this completed form to **207-772-3105**, or mail it to:

Product Development, J. Weston Walch, Publisher, P. O. Box 658, Portland, ME 04104-0658

We will send you a **FREE GIFT** as our way of thanking you for your feedback. **THANK YOU!**